Aurélio Ferreira Borges
Maria dos Anjos Cunha Silva Borges

Fish farming in amazonia is environmentally unviable

AF153824

Aurélio Ferreira Borges
Maria dos Anjos Cunha Silva Borges

Fish farming in amazonia is environmentally unviable

LAP LAMBERT Academic Publishing

Impressum / Imprint

Bibliografische Information der Deutschen Nationalbibliothek: Die Deutsche Nationalbibliothek verzeichnet diese Publikation in der Deutschen Nationalbibliografie; detaillierte bibliografische Daten sind im Internet über http://dnb.d-nb.de abrufbar.

Alle in diesem Buch genannten Marken und Produktnamen unterliegen warenzeichen-, marken- oder patentrechtlichem Schutz bzw. sind Warenzeichen oder eingetragene Warenzeichen der jeweiligen Inhaber. Die Wiedergabe von Marken, Produktnamen, Gebrauchsnamen, Handelsnamen, Warenbezeichnungen u.s.w. in diesem Werk berechtigt auch ohne besondere Kennzeichnung nicht zu der Annahme, dass solche Namen im Sinne der Warenzeichen- und Markenschutzgesetzgebung als frei zu betrachten wären und daher von jedermann benutzt werden dürften.

Bibliographic information published by the Deutsche Nationalbibliothek: The Deutsche Nationalbibliothek lists this publication in the Deutsche Nationalbibliografie; detailed bibliographic data are available in the Internet at http://dnb.d-nb.de.

Any brand names and product names mentioned in this book are subject to trademark, brand or patent protection and are trademarks or registered trademarks of their respective holders. The use of brand names, product names, common names, trade names, product descriptions etc. even without a particular marking in this work is in no way to be construed to mean that such names may be regarded as unrestricted in respect of trademark and brand protection legislation and could thus be used by anyone.

Coverbild / Cover image: www.ingimage.com

Verlag / Publisher:
LAP LAMBERT Academic Publishing
ist ein Imprint der / is a trademark of
OmniScriptum GmbH & Co. KG
Heinrich-Böcking-Str. 6-8, 66121 Saarbrücken, Deutschland / Germany
Email: info@lap-publishing.com

Herstellung: siehe letzte Seite /
Printed at: see last page
ISBN: 978-3-659-42700-8

Zugl. / Approved by: Lavras, University of Lavras Brazil, Postdoctoral Theses, 2014.

Copyright © 2015 OmniScriptum GmbH & Co. KG
Alle Rechte vorbehalten. / All rights reserved. Saarbrücken 2015

SUMMARY

FISH FARMING IN AMAZONIA IS ENVIRONMENTALLY UNVIABLE

Abstract: The research aimed to describe the performance of Environmental Management of Aquaculture at Amazonia in Brazil. This is applied in non-experimental (descriptive) research. Criteria for the classification of Environmental Management Performance on the aquaculture is below to 30%, between 30.1 and 50%, between 50.1 and 70%, between 70.1 and 90% and higher to 90.1%. The performance indices of Environmental Management obtained proved that the aquaculture is environmentally unsustainable.

Keywords: Environmental regulations. Legislation. Cronbach Alpha. Qualitative and quantitative research.

1.INTRODUCTION

According to Ministery of Fishing and Aquiculture (MPA, 2012), the world production of fish (from extractive fishery and aquiculture) attained approximately 168 million tons on 2010, representing an increase of 3% in relation to 2009. The larger producers were China with 63.5 million tons, Indonesia with 11.7 million tons, and India with 9.3 million tons and Japan with 5.2 million tons. Brazil has contributed with 0.75% (1,264,765 ton.) of the world fish production on 2010, occupying the 19th place (Table 1). Considering the countries of South America, the fish production on countries that fish at the Pacific Ocean is superior to the Brazilian production. Peru recorded a production of 4.4 million tons, followed by Chile with 3.8 million tons. In this criterion Brazil appears in the third place (Table 1).

Table 1 Total production of fish (ton) of the 19 largest producer countries.

Position	Country	2009		2010	
		Production	%	Production	%
1°	China	60,474,939	36.95	63,495,197	37.69
2°	Indonesia	9,820,818	6.00	11,662,343	6.92
3°	India	7,865,598	4.81	9,348,063	5.55
4°	Japan	5,465,155	3.34	5,292,392	3.14

3

5°	Philippines	5,083,218	3.11	5,161,720	3.06
6°	Vietnam	4,870,180	2.98	5,127,600	3.04
7°	United States	4,710,653	2.88	4,874,183	2.89
8°	Peru	6,964,446	4.26	4,354,480	2.59
9°	Russia	3,949,267	2.41	4,196,539	2.49
10°	Myanmar	3,545,186	2.17	3,914,169	2.32
11°	Chile	4,702,902	2.87	3,761,557	2.23
12°	Norway	3,486,277	2.13	3,683,302	2.19
13°	South Corea	3,201,134	1.96	3,123,204	1.85
14°	Thailand	3,287,370	2.01	3,113,321	1.85
15°	Bangladesh	2,885,864	1.76	3,035,101	1.80
16°	Malaysia	1,874,064	1.15	2,018,550	1.20
17°	Mexico	1,773,713	1.08	1,651,905	0.98
18°	Egypt	1,292,889	0.67	1,304,795	0.77
19°	Brazil	1,240,813	0.76	1,264,765	0.75

Source: adapted from MPA (2012)

According to MPA (2012) the national production of fish by Federation Unit for 2011 demonstrated that the state of Santa Catarina remained as the largest producer of fish in Brazil, with 194,866.6 tons (13.6%), followed by the states of Pará with 153,332.3 tons (10.7%) and Maranhão with 102,868.2 tons (7.2%). The states of Bahia, Rio Grande do Sul, São Paulo, Mato Grosso, Alagoas, Sergipe and Distrito Federal presented a reduction in relation to the amount produce on 2010. However, for the other states an increase was observed on the production of fish in this period.

Continental aquatic resources constitute an essential component of all terrestrial ecosystems. The generalized lack of water, the gradual destruction and aggravation of pollution of hydric resources in many regions of the world, allied to the progressive implantation of incompatible activities have increasingly required the planning and integrated management of these resources (OSTRENSKY; BORGHETTI, SOTO, 2008).

According to DA SILVA (2013), the terrestrial part of Brazilian seaside has changeable width, enlarging almost 10,800 Km throughout the coast[1] if counting the depressions on natural surfaces, and comprises an area of almost 514.000 km² along the 17 coastal states. The fishery is among the fundamental economic activities incremented throughout the Brazilian littoral.

The article 3rd, §2nd, of Decree number 23,672, of January 2nd 1934 (BRAZIL, 1934), which approved the Code of Hunting and Fishery states that "the coastal fishing is made from the coast until the distance of 12 miles counting out".

During the Second World War Brazil had supported regional actions for the constitution of a constancy range of 12 miles. However the territorial sea remained with width of 3 miles as constituted since half the century XVIV (DA COSTA, 2013). According to the author Brazil would only add the territorial sea for 6 miles by means of the Decree-Law number 44, from November 18th 1966 (BRAZIL, 1966). Furthermore, this decree still stablished a supplementary range until the extent of 12 miles of the littoral in order to prevent and reduce infractions regarding customs, taxation, health or immigration police.

According to this author this period would not persist for a long period. Immediately on April 25th 1969 the Brazilian government edited the Decree-Law number 553 (BRAZIL, 1969), also decomposing the limits of Brazilian territorial waters, what came to be 12 nautical miles adjusted from the low-water line (article 1st).

The author still added that the period with limit of 12 nautical miles for the territorial waters would be shorter than the previous one. Less than one year after, on March 25th 1970, through the Decree-Law n. 1,098 (BRAZIL, 1970) the Brazilian government determined that the "territorial sea of Brazil comprises a range of two hundred nautical miles width measured from the low-water line of the Brazilian mainland and island coastline" (article 1st). Itamaraty acquired protest reservations of some countries, with great fishery interests or conducted by positions favoring a narrow territorial waters and, in all the responses, the Ministry of External Relations reaffirmed the persuasion that there was no invigorating norm of international law, neither enshrined, nor customary, that determined to the State the maximum limit until which it can extend its territorial waters in Brazil.

Even before the beginning of the international validity of United Nations Convention on the Law of the Sea (CNUDM) which occurred on November 16th 1994, had already adequate to the parameters provided by the international treaty, with the advent of Law number 8,617 from January 4th 1993 (BRAZIL, 1993), which dispose about the territorial waters, the Brazilian contiguous zone, the Exclusive Economic Zone (ZEE) and the continental shelf, revoking the Decree-Law number 1,098/1970 and other provisions to the contrary.

1.1 The continental shelf and its relation with sea fishing

According to DA SILVA (2013), with the end of the Second World War and before the territorial waters were stipulated with new limits, the attention of Brazil for the Law of the sea turned to an over again unknown area: the continental shelf, still called underwater shelf. This author stated that the Brazilian government, as well as other governments had made on Latin America formerly, equally affirmed that the continental shelf was the complementary part of the national territory. This occurred by means of the Decree number 28,840, from November 8th 1950 (BRAZIL, 1950), in which "is expressly recognized that the underwater shelf in the part corresponding to the continental and island territory of Brazil is integrated in this territory under exclusive jurisdiction and domain of the Federal Union" (article 1st).

The author still added that, as stressed by the article 3rd, the inclusion of continental shelf did not decomposed the regime of territorial waters, which

6

remained as 3 miles: "continue valid the norms about the navigation in waters overlapping the previously mentioned shelf, without prejudice of the posteriorly stablished, especially about fishing in this region". Continuing the expansion of territorial waters from 3 to 6 nautical miles, the Decree n. 62,837 from June 6th 1968 (BRAZIL, 1968) was edited, which organized the exploration and research on the Brazilian shelf, on territorial and inland waters, and considered the underwater shelf a part of the national territory according to the Federal Constitution, and understanding it as "the seabed and subsoil of submarine areas adjacent to the coast, but situated outside the territorial waters until 200 meters depth", then understanding that the "expressions 'underwater shelf', 'continental shelf' and underwater continental shelf' are equivalent" (art. 3rd).

1.2 Natural capital and ecosystem services

According to ANDRADE (2011), the guiding principle is that the treatment given to the environment is mostly reductionist and overcoming it must necessarily pass through transverse approaches. The catenation of the described ideas converges to the need to build this new structure of analysis of the called Ecosystems Economy.

The author still added that the economic system considered as an alive and complex organism does not act independently of the natural system that supports it. Contrarily, it interacts with the environment extracting natural resources (structural components of the ecosystems) and returning wastes in a process of absorption of matter and low entropy energy, and discards the residues of high entropy. The impacts generated over the environment are function of the scale (size and dimension) of the economic system and the way the economic growth occurs (the way the system expands).

In the perspective of the author, the term "natural capital" was firstly used as metaphor to refer to natural resources available to the men. However, only in the end of twentieth century the term was no longer used and then was a way to claim for attention to the problem of natural resources depletion, thus consisting a formal and technical concept used together with definitions of other types of capital[1].

For the author, the concept of natural capital considers all the flows of tangible (natural resources) and intangible (ecosystem services) benefits from all the natural resources, which are direct and indirectly appropriable for men. This wider concept confers to the natural capital a multidimensional character, in which ecological, economic and sociocultural dimensions are related and interact to promote the human welfare.

According to VEIGA (2010), the synthetic index that measures the human welfare was proposed by researchers of Yale and Columbia, builders of an Environmental Sustainability Index (ESI) and Environmental Performance Index (EPI). The first contains 76 variables that cover five dimensions. The second aggregates the same 76 variables on 21 intermediate indicators. For the author, even these are reasonable manners to group a great number of information and serve as invitation to attract more attention to some of its components, all these types of exercises are highly precarious under an strictly statistical point of view.

The author adds that the Ecological Footprint is an indicator that intends to show how much of the regenerative capacity of the biosphere is being used by human activities (consumption). Initially proposed by WACKERNAGEL & REES (1995), this indicator has been promoted by Global Footprint Network, by Redefining Progress and also WWF, which publish the updates in the annual Living Planet Report. However, the apparent simplicity of Ecological Footprint also hides serial technical

8

problems. For example, the biocapacity of a cultivated area is checked by the observed yield, when it should be checked by the yield that would allow remaining constant the fertility of this soil in the future, it means, its sustainable yield. The same occurs with the evaluation of biocapacity on pastures. Thus, nationally the ecological deficit of these lands will always be equivalent to the commercial deficit of the sector. And in world scale there will never be ecological deficit or surplus relative to the agriculture.

Under this author perspective, there are similar deficiencies on methods of calculus regarding the built areas, forest areas and fishing. However, more important than criticizing them is to claim for attention to the fact that on the footprint conception there is a subjacent or intrinsic weighing. It would be reasonable admitting that the relative importance of forests represents only 9% and fishing 3%? Lastly, in this sense it is admitted that a substitution of forests by cultivated lands would increase the available biocapacity, then relieving the ecological deficit, what does not make sense.

1.3 On the relation of pisciculture and environment in Rondonia

According to BRAZIL (2008) on Article 5th it is declared of social and economic interest the fishing with activation purposes in use of Permanent Preservation Area of Rondonia, already anthropic, attended the requirements established in this Law.

§ 1st – The construction of water reservoirs, dams, weirs and tanks used for the implantation of fishing activities may be licensed on water courses with mean flow of 1 m³ (one cubic meter) per second.

§ 2nd – For the construction of water reservoirs, dams, weirs and tanks used for the implantation of fishing activities on water courses with mean flow higher than 1 m³ (one cubic meter) per second, the interested will require to the State Secretary of Environmental Development of Rondonia (SEDAM) an special license.

According to the author on Article 6th the implantation of fishing activity will be authorized on Permanent Preservation Area (APP) that irregularly were subtracted to be substituted by pastures and other agricultural activities when the applicant:

I – prove the indispensability of the intervention of the Permanent Preservation Area (APP) for the total financial-economic feasibility of the enterprise.

On Article 7th it is established that the artificial reproduction of native species of fish in Rondonia destined to the production of pure fries must occur on laboratories properly licensed for this purpose by the competent organ:

Single paragraph – Fries acquired from other states or countries must be accompanied by the sanitary inspection report.

According to the author on Article 10th the environmental licensing of fishing will be processed on the SEDAM of Rondonia on the modalities Previous License, Installation License and Operation License, and the applicant must indicate the classification of the activity on the terms of Articles 3rd and 4th of this law, presenting the technical project with the constant specification on Terms of Reference for the aquaculture activity issued by the environmental state agency.

The author still adds that on Article 17 the values of expedition rates of Previous License (LP), Installation License (LI)) and Operation License (LO) for aquaculture activities in the state of Rondonia will be calculated based on the Fiscal Standard Unit of Rondonia according to the Creation System:

I – Enterprises with until 02 (two) hectares for the Systems of Creation I, II and III and of until 50 (fifty) cubic meters of water for the System of Creation

IV are free from rates. Since these systems are explored by small farmers besides being considered of low environmental impact, only the presentation of the Environmental Control Report (RCA) for licensing will be required, which must be made by properly accredited professionals or entities.

2. OBJECTIVES

2.1. General
Describe the performance of Environmental Management of Aquaculture at Amazonia in Brazil.

2.2. Specific

a) Characterize the social and environmental standards on properties of the study area.
b) Identify standards of food and health safety on fish cultivations of the study area.
c) Determine the value of alpha Cronbach coefficient of the questionnaire used in the research.

3. RESEARCH PROBLEM

OSTRENSKY, BORGHETTI and SOTO (2008) have verified several problems on the fish sector when interviewing Brazilian fish producers. The greatest problem was related to environmental issues.

According to ROCHA et al. (2013), although the consumption of fish is still low in Brazil, the national commercial balance of fish is in deficit since 2006 both in monetary terms and in volume sold. Data estimated by the Ministry of Fisheries and Aquaculture (BRAZIL, 2010) showed that the importation of fish and byproducts attained US$ 1,262,888,212 (349,529,158 Kg), while exportation of national product attained only US$ 271,193,147 (42,263,415 Kg), what represents a deficit of US$ 991 million (307,265,743 Kg) and elevation of 32.5% in relation to the deficit counted on 2010, which corresponded to approximately US$ 748 million.

According to ROUTLEDGE et al., (2013), the main problems faced by the Brazilian aquaculture are: a) the lack of public politics for the activity development; b) the lack of training and technical qualification of the aquaculture productive chain; c) the difficulty to access credit for investment and funding in aquaculture; d) the need to improve the competitiveness of aquaculture in small scale; e) the need to make feasible the processing of products derived from aquaculture in industrial scale; f) the need to create a national control system of aquaculture health; g) the need to conquer new markets; h) the need to streamline the regulation of aquaculture enterprises on physical spaces of water courses of the Union domain; i) the need to survey and divulgate basic sector information.

Diffuse attempts to foment the aquaculture were experienced by the pisciculture activity, which driven by the excellent natural conditions that Brazil present, by the effort of selfless producers, professionals and dreamers of several areas, achieved the stage in which is nowadays found. But it is time to cause a professionalism and organization outbreak, preferentially in all the productive chain that composes the activity.

There are several isolate actions being adopted in Brazil that aim to order the pisciculture productive chain. Representatives of the productive sector are worried with the reduction of costs, increase of quality and competitiveness of their products, with the environment and rational use of natural resources.

The present research aimed to evaluate the performance of Environmental Management of pisciculture in the municipalities of Colorado do Oeste and Ariquemes located in the state of Rondonia, Western Amazon of Brazil.

The research problem arose from the question: how the performance of Environmental Management of pisciculture in the municipalities of Colorado do Oeste and Ariquemes is classified when considering three main issues: a) social and legal standards; b) environmental standards and c) food and health safety standards.

4. THEORETICAL FRAMEWORK

According to OSTRENSKY, BORGHETTI and SOTO (2008), the Federal Law number 9,433/1997 (BRAZIL, 1997) linked issues of hydric resources to environmental issues. In this sense the water may not be considered raw material of the productive system of agriculture and livestock. Thus, the State Policy of Hydric Resources that attempted to complement the Environmental National Policy through the Law number 6,938/1981 (BRAZIL, 1981) did not intend to do the same with the National Agricultural Policy by means of the Law number 8,171/1991 (BRAZIL, 1991). This conflict of productive and preservationists interests that frequently already occurs in the practice ends up creating antagonisms on legal instruments of the same hierarchy (Federal Laws) in which one acts as developer and other as regulator.

Authors mention that the complex legislation requires from the fish farmer the obtainment of record, licenses, grants, transfers, which mostly are costly and complex processes. The Federal and State uncertainties about responsibilities and rules for the emission of such documents increase the legal fragility of such enterprises. These events worsen the situation of fish farmers who pleads for funding for their crops, because it makes them ineligible to the credit since they do not have the documents.

The environmental licensing is a legal obligation previous to the installation of any enterprise or activity potentially polluting or degrading the environment and has as one of its more expressive characteristics the social participation on decision makings by means of Public Audiences as part of the process.

According to the Federal Constitution of 1988 (BRAZIL, 1988) is "common competence of the Union, States, Federal District and municipalities protect and combat the pollution in any form" (item VI of Art. 23) and "preserve the forests, fauna and flora" (item VII of Art. 23). However, considering that the ideal is the unique and non-simultaneous environmental licensing by the three spheres provided by the Constitution, the National Congress needs urgently to legislate over the Complementary Law provided on the single paragraph of the Article 23 of Federal Constitution explaining the means of cooperation among the federative entities for the environmental licensing. This is a problem that affects the country as a whole and not only the pisciculture, because any activity liable to licensing ends up facing this problem (OSTRENSKY; BORGHETTI, SOTO, 2008).

According to CLARK (1992), sustainable development is an intelligent and responsible way to use natural resources without harming the economic value of the natural source for the future generations. According to NEW (2003) the sustainable pisciculture and responsible aquaculture are generally used as synonyms, but he prefers to use responsible due to the great importance that this word implies for the sustainability. Thus, responsible aquaculture is to make a profitable pisciculture, but with awareness.

NOGUEIRO (2008) stated that the methods of environmental

14

management are elements whereby organizations may improve their performances. According to him, an eco-efficient procedure reduces the use of resources and prevents the production of wastes, thus acquiring significant savings.

The concept of public environmental management highlights the conciliator aspect of State regarding environmental issues.

> Public environmental management is a process of mediation between interests and conflicts among social actors that act over the physic-natural and built means. The mediation process defines and redefines continuously the way that different social actors alter the quality of environment through their practices, and also how they distribute the costs and benefits caused by their actions on the society (FLORIANO, 2007, p. 2).

According to this author, with the guidelines of Agenda 21 of the Statement of Rio on 1992, it is verified that public policies of environmental management must have as objective not only the managements of resources to protect the natural environment, but mainly to serve as guidance on the solution of social conflicts that involve environmental issues in view of the social welfare and conservation of resources for future generations.

Environmental management (GA) is the supervision of economic and social activities in order to rationally employ the natural resources, renewable or not. According to Nogueiro (2008), environmental management must be directed the use of practices that cover the permanence and preservation of biodiversity, recycling of raw-material, the decrease of environmental impact of human activities and accomplishment of environmental legislation about natural resources. Nogueiro (2008) also completes the structure of knowledge regarding the techniques to recover degraded areas, reforestation techniques, procedures for the sustainable exploration of natural resources and the sketch of risks and environmental impacts for the estimative of new enterprises or improvement of productive and educational activities.

4.1. Code of better management practices of for aquaculture

According to SOTO et al., (2008), aquaculture growth international invariably includes (with differences amongst regions and economies) the growth of cultivated areas, higher density of aquaculture installations and of renowned individuals, and use of feed resources produced external of the immediate area. Potential negative effects of aquaculture on the ecosystem often include: (i) cumulative demands on fisheries for fish meal/oil, major constituents of carnivorous/omnivorous species feeds, (ii) nutrient and organic supplementation of recipient waters resulting in build-up of anoxic sediments and modifying benthic communities (iii) eutrophication of lakes or coastal zones, (iv) restructuring of biological and/or social environments, (v) release of chemicals used to control water conditions and diseases (vi) competition for, and in some cases depletion of resources (e.g. water) and (vii) negative effects from escaped farmed organisms, often more relevant when exotics. But on the other hand aquaculture can have positive effects on the ecosystem, for example by providing the seed for re stocking of endangered or over exploited aquatic populations. Often as well, aquaculture is negatively affected by other human activities such as contamination of water ways by agriculture or industrial activities.

In the perspective of authors, in an attempt to control insufficient progresses countries worldwide have employed a large number of aquaculture guidelines. These have varied from general rules such as prohibition the utilization of mangroves for aquaculture practices to very specific regulations such as the establishing of maximum production per area,

16

regulations for disease control, use of drugs, etc. However, these regulations, neither on their own or taken together, provide a comprehensive context ensuring a sustainable use of aquatic environments. That will happen when aqua farming is treated as an integral process within the ecosystem.

For authors, development of innovative technologies has made production more efficient and facilitated increase. But often the regulations in place cannot guarantee sustainability, especially as most of them focus on the individual farmer and do not consider additive (cumulative) or synergistic effects of many farms on a particular area. Instantaneously, farmers' economic appraisal tends to have a narrow (short-term) view, focused on the more immediate production consequences. Such considerations do not include the medium and long term revenues and costs that may be imposed to the farming activity itself and on the rest of the society in the form of a reduced supply of ecosystem goods and services.

According to SOTO et al., (2008), an ecosystem approach to aquaculture (EAA) to produce an agreed set of concepts, scales and some management measures for the implementation of an EAA (Table 2).

Table 2 Summary of guiding principles, scales and major issues under each.

Principles	1	2	3
Scales	Aquaculture should be developed in the context of ecosystem functions and services (including biodiversity) with no degradation beyond their resilience.	Aquaculture should improve human wellbeing and equity for all relevant stakeholders.	Aquaculture should be developed in the context of other sectors, policies and goals.
Farm	Better/best management practices implemented at this scale. Large intensive farms may significantly alter local/site ecosystem functions. Farmed species escapes and diseases take place and can be controlled at this scale.	Returns to local farmer are often unfair. Aquaculture can offer family improvement options and employment opportunities. Working conditions are not always adequate.	Use of on-site and immediate surrounding resources more common in Asian countries (e.g. integrated agriculture-aquaculture).

	Integrated aquaculture can be an opportunity for mitigation of environmental impacts.	Food safety can often be a concern at this scale especially for small farmers.	
Watershed/ zone	Environmental effects of clusters of farms are rarely being evaluated. Limited knowledge to define ecosystem resilience capacity. Diseases and establishment of alien species take place at this scale and could be prevented, mitigated.	Unplanned/unregulated aquaculture activities could increase inequity. Often different stakeholders have different abilities/ opportunities to access resources and benefits from aquaculture. Increasing equity and well-being simultaneously will not always be possible. Transfer of benefits from regional, national and other scales should get to the local scale. Local scale initiatives promoting well-being and equity are often ignored.	Lack of support and regulations for integrated aquaculture and multitrophic aquaculture. Local scale activities/initiatives most often are not subsidiary to the wider context of watershed, coastal zone management policies and programmes. Integration between different sectors are not been facilitated within the ecosystem perspective. Geographical remit of aquaculture development. authorities' remit (i.e. administrative boundaries) often do not include watershed boundaries.
Global	Increasing pressure on small pelagic fisheries for fishmeal to feed aquaculture. Unknown biochemical consequences of N, P, C transport among regions partially driven by aquaculture. Climatic change affecting aquaculture development in the ecosystem context.	Improving the well-being of relevant stakeholders within the context of trans-national aspects of production, and markets is a challenge and an opportunity. Food safety globally enforced due to global markets. Development of global opportunities can compromise regional and local opportunities.	Fish and aquatic proteins are increasing in world diets, and aquaculture is rapidly increasing its relevance. Competition with other food and energy sectors for vegetable proteins (feeds) is increasing. Competition for freshwater use with other food sectors will increase.

Source: adapted from Soto et al., (2008).

Making sure to promote independent research to facilitate compliance of the 3 principles at the farm scale and beyond is at the core of EAA. Of particular relevance is to promote research contributing to the understanding and planning of the production process within the ecosystem framework. At all scales research needs to be interdisciplinary and multidisciplinary and long-term, there is also the need for research on governance which includes/considers a balanced ecosystem. Governments should have a more inclusive process for decision-making regarding aquaculture and appropriately devolved power at the local scale. Of great relevance is the development of Simulation Models as decision tools at different scales. Research on valuation of ecosystem services which may be undermined by aquaculture are most important in order to properly plan location of farms and aquaculture zones, mitigation measures, maximum production allowances.

For the farm scale research should emphasis on increasing tools to calculate externalities of inputs and outputs, to estimate carrying capacity for individual farms, and tools and technologies for improving the feeding process and conversion ratios. It is also very important to promote permanent and proactive research on new species and strains offering enough information for the selection of the right species based on ecosystem functions and market demands, considering species requirements and ecological/nutritional efficiency. Aquaculture rapid development and risks for sudden crash should be avoided/prevented ensuring continuity in following-up in regulation management and reinforcement processes, irrespectively of fluctuations in governments, and authorities in charge (SOTO et al., 2008).

According these authors, at the watershed scale research should measure the inspection of the most appropriate species to farm, including potentially new species, while closing the life cycles of species of interest could also be important for the diversification of aquaculture within a watershed or at least to keep a wide number of candidate species as an insurance for the sustainability of the sector in the watershed. At this scale, practical research should also cover health management and biosecurity.

The promotion of integrated aquaculture including integrated multitrophic aquaculture (IMTA) is a logical way to insert aquaculture in an aqua-system or aqua-agro system where there is proper recycling and full utilization of resources and energy while diminishing risks associated to by-products and increasing productivity of the sites. However, a proper valuation of the externalities in monocultures needs be consider in order to enhance integrated aquaculture (SOTO et al., 2008).

For these authors, the application of EAA to be successful, stakeholders must understand and accept the need for this more integrative approach to aquaculture assembly. This will require a proactive effort by management agencies particularly ensuring effective and appropriate training for all staff having to deal with the changes required for EAA. Scientists and management authorities will need to recognize the value of the knowledge of fisher folk and aqua farmers, their representatives and communities (particularly regarding the ecosystem). They must also recognize that with the ever-broadening range of stakeholders under EAA, the potential differences in capacity to participate in management will also increase which, if uncorrected, will lead to unbalanced and poor decisions.

At the watershed scale it may be essential to simplify integration amongst farmers, and amongst farmer's associations (e.g. mussel farmers and fish farmers) also simplifying integration with fisheries and fisher folk, with agriculture, recreation, urban and industrial activities and stakeholders. This should also include research, common resource management, education; etc. Obviously, facilitating decentralization of management at the watershed level can be an important step.

At the global scale it is significant to endorse generation of evidence with clearness to aquaculture and to other sectors and consumers on the advantages of such integration. It could also be possible to contribute/promote the development of eco labels and or other certification tools to acknowledge integration and the implementation of EAA (SOTO et al., 2008).

The Food and Agriculture Organization of the United Nations (FAO) establishes since 1995 the principles and international standards in order to guarantee the effective conservation and management of aquatic resources with respect to the ecosystems and biodiversity. The Code of Conduct for Responsible Fisheries (CCPR) (FAO, 1995; 1999b) was formulated to be interpreted and applied according to the international laws based on the Convention of United Nations referring to the sea law of 1982, also on the Declaration of Cancun, Mexico from 1992 and on the Declaration of Rio de Janeiro from 1992 about development and the environment, especially chapter 17 of Agenda 21.

The article number 9 of CCRF called Development of Aquaculture has as objective to serve as general guidelines for the interested on carrying on initiatives in support of a sustainable development of pisciculture. According to FAP (1999a), the establishment of an environment adequate to the sustainable development of pisciculture is responsibility of who occupies government positions and its institutions, the specialists on natural sciences and sociologies, the media, financial institutions, associations of social and private sector, as well as of aquiculture producers, manufacturers and suppliers of inputs, manufacturers and traders of aquaculture products. The commitment with understanding and equity and a responsible attitude on consultations and negotiations among countries or regions will help the sustainable development of aquaculture.

The use of reservoirs for multiple uses on Brazil was established by the National Policy of Hydric Resources on 1997 with the Law number 9,433 (BRAZIL, 1997). This legal scenario was complemented by the Decree number 4,895 from 2003 (BRAZIL, 2003) and the Interministerial Normative Instruction number 6 from 2004, which regulate the use of Brazilian waters and public spaces, it means, the "federal waters" for the practice of aquaculture. Furthermore, the Interministerial Normative Instruction number 7 from 2005 (BRAZIL, 2005) specifies that until 1% of the surface area of federal waters are available for aquaculture purposes, what corresponds to approximately 55,000 hectares.

In the perspective of authors, the Conservation Units (CUs) are territorial spaces created and instituted by the National System of Conservation Unit the Law number 9,985 of 2000 (BRAZIL, 2000) which objective is the conservation of natural resources. CUs are divided in two groups, of integral protection and sustainable use. The CUs of integral protection do not allow the compatibility among production activities and consequently its limits are considered as improper area for the demarcation of parks for aquaculture.

Authors still add that in the last centuries the interventions on aquatic environments has been particularly dramatic, especially those caused by the increase of population, industrialization and the need to generate hydropower. Nowadays the rate in which reservoirs are built is next to one per day, existing approximately 50,000 large dams built along the world, what together have the capacity to store about 6,500 Km^3, occupying an area equivalent to the French territory.

According to BRITTON and ORSI (2012), Brazil has a highly diverse freshwater fish fauna and their freshwaters provide valuable provisioning ecosystem services in aquaculture and sport angling, especially in the developed regions in the south. Non-native fish now comprise a substantial proportion of the total aquaculture production and value, contributing at least US$ 250 million in 2008 (63% of the total value of freshwater fish

22

aquaculture) according to the Fish and Agriculture Organization. Much of this aquaculture activity is centered in Central and Southern Brazil, such as impounded sections of the upper River Paraná. The non-native fishes used tend to feed at relatively low trophic levels, with the most prominently species being *Cyprinus carpio* and *Oreochromis niloticus*. Ecological risk assessment suggests these species are potentially highly invasive and deleterious to the native fish diversity of invaded water bodies.

Authors still add that fishes introduced for the creation of sport fisheries tend feed higher trophic levels through piscivory, such as the peacock basses (*Cichla* species) from Amazonia. Their introductions have generally resulted in establishment and invasion, which tends to be followed by significant and rapid declines in native fish diversity as a consequence of increased predation pressure. Thus, whilst non-native fish in the upper Paraná River support provisioning ecosystem services of substantial economic value, the principal species used represent high risks to fish diversity and conservation. It is recommended local management should concentrate on reducing these risks through use of more appropriate species in these ecosystem services, with these decisions derived using risk assessment and precautionary principles.

According to BUENO et al. (2013), the Brazilian Government has been promoting studies on the zoning and demarcation of aquaculture parks dedicated to the production of fish in net-cages in the large public reservoirs of the country.

Authors still add that the methodology engaged for the zoning of these aquaculture parks involves of the execution in three consecutive stages of multidisciplinary evaluations for their social, environmental and economical characterization (global, regional and local). The determination of the studies includes the creation of thematic maps and situations of environmental

23

models that facilitate the process of understanding the regional peculiarities and taking the decisions to identify the most appropriate areas for the installation of the aquaculture parks. The submission of this mechanism for zoning the reservoirs will assist a more effective development of investments and efforts, both by the government and by private initiative, in the aquaculture activity. With the establishment of aquaculture parks along the lines of an ecosystemic aquaculture, Brazil has the budding to become one of the predominant producers of fish in net-cages in the world.

In the perspective of the authors, one of the principal modalities of aquaculture being developed in Brazil is the rearing of freshwater fish, especially the Nile tilapia (*Oreochromis niloticus*) and, more recently the rearing of Tambaqui (*Colossoma macropomum*) and pintado (*Pseudoplatystoma spp.*) in net-cages systems installed in large reservoirs.

The utilization of the reservoirs for multiple uses in Brazil, aquaculture among them, was stablished by the National Policy of Hydric Resources in 1997 with the Law number 9,433 (BRAZIL, 1997). This legal framework was completed by Decree number 4,895 by 2003 (BRAZIL, 2003).

According to PELICICE et al. (2014), as Brazil undergoes rapid economic growth, short-sighted political decisions can threaten biological diversity and ecosystem services. Recently, the Brazilian Congress proposed a law to allow the rearing of nonnative fish in aquaculture cages in any hydroelectric reservoir of the country. This initiative may "naturalize by decree" some of the worst invasive species in the world (e.g., carps and tilapias) as a means of developing inland aquaculture and economy. The spread of aquaculture facilities will create opportunities for fish invasions to occur throughout the country, with the risk of damaging native biodiversity, ecosystem services, and environmental quality on a continental scale.

In the perspective of the authors, the proposal ignores ecological

24

theory, historical and/or empirical data concerning fish invasion, including dispersal, establishment, propagule pressure, invasiveness and invasibility, and all the negative consequences that may follow the invasion and establishment of nonnative organisms. This situation inspires reflection about the future of tropical biodiversity worldwide, particularly because Brazil, like many other developing countries, possesses a remarkable diversity of fish and other freshwater organisms yet has taken some political measures that are in conflict with important conservation issues.

Authors still add that Brazil is a fitting example for reflection and discussions about the challenges imposed by biological invasions and other conservation issues in developing countries. As the country undergoes rapid economic growth, several political decisions have seriously threatened its biological diversity and ecosystem services. Amid this wave of policies that prioritize fast economic growth over sustainability, the Brazilian Congress has proposed another controversial law (Law Project 5,989/2009) that want to allow the rearing of nonnative fish in aquaculture cages to be installed in hydroelectric reservoirs.

The belief that nonnative fish in aquaculture cages may increase food supply and decrease the demand for natural stocks, promoting economic growth and reducing poverty, has led developing countries and international agencies (i.e., the World Bank) to undertake major investments in these topics. Although the spread of aquaculture with nonnative species will certainly foster economy, especially in emergent countries, this system also will create opportunities for massive fish invasions, with the risk of promoting the loss of native biodiversity, ecosystem services, and environmental quality (PELICICE et al., 2014).

According these authors, the first version of the Proposed Law (henceforth mentioned to as Law Project) dates from 2009. The central objective is to allow the rearing of nonnative fish in aquaculture cages installed in the reservoirs of large hydroelectric dams. Because the introduction of nonnative species is prohibited by law in Brazil, the PL intends to remove the legal obstacles for fish production. The original version proposed the rearing of carps (i.e., *Aristichthys nobilis, Ctenopharyngodon idella, Cyprinus carpio,* and *Hypophthalmichthys molitrix*) and Nile tilapia (*Oreochromis niloticus*), which would be "naturalized by decree" to attain a native status. Naturalization in Law Project means that these nonnative fish will be considered legally native in Brazil, based on the argument that carps and tilapias have been registered or established populations in some Brazilian inland water bodies.

Following several revisions, the new version of the law does not indicate the species that will be legalized, but that decision will be made by the Brazilian Aquaculture and Fisheries Secretary (Ministry of Fisheries and Aquaculture), a government agency concerned with the production of fish and other aquatic organisms. At present, the proposal has been approved by three commissions of the Brazilian Congress and remains under analysis pending final approval (i.e., Brazilian Senate and the President).

The original version of the Law Project included nonnative species with high invasiveness and recognized invasion history (carps and tilapias), but the most recent version does not indicate the species that will be naturalized. Considering the current trends in Brazilian aquaculture, we expect intense pressure to allow the use of catfishes (e.g., *Clarias gariepinus* and *Ictalurus punctatus*) and especially, tilapias (e.g., *O. niloticus, Tilapia rendalli*). It has occurred in the aquaculture parks and aquatic farms created in some regions of the Brazil, which acquired specific licenses to raise tilapia in confinement (e.g., Furnas Reservoir, Grande River (AZEVEDO-SANTOS et al., 2011; PELICICE et al., 2014).

26

Regardless of the chosen species, the possible widespread release of nonnative fish into Brazilian ecosystems is a major concern. There is no safe confinement in aquaculture, and such endeavors have been considered a main vector for the release of nonnative fishes worldwide. Escapes are inevitable, and cage aquaculture may create a constant and intensive flow of nonnative propagules into the wild (AZEVEDO-SANTOS et al., 2011). Numerous scientific publications show that negative effects follow the invasion and establishment of nonnative fish Specifically, in Brazil, there are studies reporting eutrophication, species loss, changes in community structure, faunal homogenization, the introduction of parasites, and alterations in fishery systems (e.g., PELICICE & AGOSTINHO, 2009). These problems are recognized by scientists and include the effects caused by tilapias. Yet, these concerns were neglected in discussions of Law Project, even though Brazil has experienced a long history of fish invasion via aquaculture and other activities (BRITTON & ORSI, 2012).

Furthermore, Law Project ignored other general information concerning fish invasion. For example, it ignored that tilapias and carps are among the worst invasive species in the world and that these organisms present high invasiveness, disturbance potential, negative impacts and invasion history in Brazilian ecosystems, as shown by historical and empirical data, risk analysis and ecological modelling. In addition, Law Project ignored the evidence that tilapia, carp and many other nonnative species are not established in most Brazilian river basins and in many reservoirs; for example, nonnative fish species are largely absent in the Amazon Basin. In this context, the expansion of aquaculture activities will accelerate fish introduction, establishment and dispersion events across the country. Furthermore, at sites where nonnative fish are established, Law Project (PL) ignored the potential effects of propagule pressure on the demography of the invader, which, in turn, determines the consequences and dynamics of invasion. Moreover, cages will be installed in reservoirs, environments that facilitate invasions: once in the reservoir, fish may reach contiguous areas, including reserves and protected areas. Lastly, the introduction of nonnative pathogens or parasites is a real possibility, particularly because cages remain in close contact with the surrounding environment. The spread of new parasites may harm the native biota and affect the natural dynamics of diseases (PELICICE et al., 2014).

According to these authors, if sanctioned, this law may trigger numerous invasion events that threaten resident biodiversity and damage ecosystem functions/services. We emphasize that fish invasion and its related consequences due to this specific law will not be restricted to Brazil: hydroelectric reservoirs are found in all large basins of the country (there are more than 700 large dams), several of these basins are shared with other countries (e.g., the Amazon and Paraná Rivers, Pantanal), and fish, obviously, do not recognize political borders.

The Law Project states that natural fish stocks of most Brazilian rivers are depleted and that fish production must increase. Indeed, South American freshwater ecosystems have been deeply modified by human. In addition, recent studies reported biotic homogenization and shifts in the longitudinal and latitudinal body size patterns (VITULE et al. 2012). These losses and changes are the result of several types of human disturbance, including habitat loss, pollution, overfishing, biological invasions and, in particular, river damming. Thus, in this scenario of decreasing biodiversity and fishery yield, aquaculture is being proposed as a tool to enhance economy, fish production, and ecosystem services.

Promoting the aquaculture of nonnative species as a conservation action, however, is an equivocal approach to mitigate or compensate for

28

environmental problems. Human disturbances will not disappear with the incentive for aquaculture, an activity that risk complicating the situation via additional negative impacts, including stressors that go beyond the invasion issue (e.g., eutrophication, visual and chemical pollution). Effective actions to conserve or restore aquatic diversity should consider the establishment and police of protected areas, preserve and restore natural hydrological regimes, re-evaluate the national hydroelectric plan and, obviously, prevent the introduction and establishment of nonnative species. In this sense, the conservation of fish diversity in Brazil and South America, as well in other developing and megadiverse regions of the world, demands urgent measures, particularly because all large rivers are regulated by hydroelectric dams or will be in coming years. In this scenario of profound ecological change, PL is ineffective as a solution to environmental problems, dangerous for Neotropical fish diversity and beneficial only to the production sector (i.e., during a short period), (PELICICE et al., 2014).

4.2. Conservation, government and fishery production

The Brazilian Government has repeatedly announced that fisheries and aquaculture production must increase in the country and that political measures will be taken accordingly (e.g., US$ 2 billion will be invested through 2014. Paralleling this support, Law Project has been approved by three commissions of the Brazilian Congress and is steadily proceeding towards final approval, in other words, is in the end of process. Because it encourages the use of nonnative fish to foster the development of aquaculture, the position of the Brazilian Government is, indeed, a paradox. Many developed countries have spent millions of dollars on programmes to prevent, control and eradicate nonnative species. Furthermore, Brazil is a signatory of the Convention on Biological Diversity and must, therefore, engage in avoiding new introductions and controlling/eradicating nonnative species (article 8h; UNITED NATIONS, 1992). A more precautionary solution would naturally use the best available knowledge to evaluate long-term negative or impeditive effects of planned introductions (STRAYER, 2012), especially because the issue is not trivial. If Law Project is approved, therefore, Brazil will set an official fishery program with low environmental responsibility, security and sustainability (PELICICE et al., 2014).

We do not deny the need/opportunity for aquaculture in Brazil, nor do we disagree that fishery production may become a means to alleviate poverty in the country. Instead, our view is that political decisions need to be founded on a balance among environmental, economic and cultural principles and that effective production solutions should be achieved through real innovation. Better political decisions would naturally look towards strategic planning to ensure that aquaculture is developed as a network across the entire landscape and society rather than as a series of isolated and/or short-sighted projects with no commitment to regional biological diversity and future generations. We emphasize that Brazil has a megadiverse native fish fauna, it is the ninth largest producer of freshwater fish through aquaculture, but production is almost entirely based on nonnative species (>80% of 94,844 mt), (VITULE et al. 2009).

According to PELICICE et al. (2014), this picture is again a paradox and more curious if compared with Myanmar, for example, a much smaller country with a poorer fish diversity, in which total production is similar to Brazil (93.948 mt), but totally based on native species (CASAL, 2006). In this sense, we believe that Brazil's fish diversity must be appreciated: it may allow a variety of cultivation techniques and market options, involving an array of regional/local opportunities. There is a number of native species already in use (e.g., *Rhamdia quelen, Piaractus mesopotamicus, Colossoma macropomum, Arapaima gigas*, species of the genus *Pseudoplatystoma* and *Brycon.*

In this sense, Brazil's fish diversity must be appreciated: it may allow a variety of cultivation techniques and market options, involving an array of regional/local opportunities. There is a number of native species already in use (e.g., *Rhamdia quelen, Piaractus mesopotamicus, Colossoma macropomum, Arapaima gigas*, species of the genus *Pseudoplatystoma* and *Brycon*, and more research should be destined to improve, enhance and promote the use of local/regional species through the aquaculture chain (VITULE, 2009).

According these authors, regionalization is an significant issue, for the reason that Brazil has a continental extension and its basins, sub-basins and localities present particular fish species and accumulations. In this perspective, research must pursue ways to appliance sustainable aquaculture according to regional or even local particularities (i.e., species selection, genetic structure, methods of production, market aspects), "small is beautiful," in other words, local small-scale aquaculture may be a good sustainable alternative. There are good examples coming from participatory management in the Amazon basin, in which local people have exploited and managed local resources for decades (e.g., pirarucu, *Arapaima gigas* in natural reserves; tambaqui, *Colossoma macropomum* in tanks; or ornamental fishes collected in the Rio Negro basin). These systems are obviously less productive and profitable in the short term, but they are more sustainable and environmental friendly than tilapia aquaculture or any other intensive system. Such differences cannot be ignored if we ask about the real goals of the fishery program, particularly if incentives are destined to develop small-scale aquaculture (Assuage poverty? Source of protein? Family income?) or to

31

satisfy large-scale corporations (Article of trade production? Exportation?).

Is believed that more information should precede any attempt to increase fishery production in Brazil. The fishery issue includes many social, technical, efficient and environmental mechanisms, so political force alone cannot deal with such a complex system. Operative plans should trust on basic ecological/fishery data, such as the status of wild stocks, the capacity of inland ecosystems to sustain an increasing fishing struggle, the carrying capacity of reservoirs to implement cage aquaculture, and all potential negative impacts that may succeed (e.g., pollution). In addition, strives to increase aquaculture productivity should be accompanied by improvements in the production system as a whole. It would include safe confinement, quarantine procedures, waste treatment, use of space and technical orientation to name a few. Seeing that government incentives concern especially with procedures to increase fish production, is supposed that conflicts in the use of freshwater resources will aggravate in the near future. In this logic, to preserve ecosystem processes and services in the long term, an ecosystem approach, at the basin scale, would be a promising alternative to manage freshwater resources and progress sustainable aquaculture (UNITED NATIONS, 1992; PELICICE et al., 2014).

The incentive to promote aquaculture with nonnative species (together with other recent administrative departures) favors especially the private production of commodities, with the appropriation of natural capital and the consequent loss of natural resources. In fact, the false dichotomies termed "economy versus ecology" or "environment versus economic growth," whereby the maintenance of natural ecosystems is viewed as an impediment to social and economic well-being, are constantly cited in political debates about these issues (e.g., the official plan to accelerate development, the Project for Economic Growth Acceleration to Brazil (PAC), held by Brazilian government. This scenario is not restricted to Brazil, considering that the uncritical use of nonnative species to achieve short-term efficient gain has been recorded around the world (CASAL, 2006; Simberloff et al., 2013; Pelicice et al., 2014), especially in emerging/developing countries that rely on aquaculture. In case of Brazil, however, this equivocal stance is trivializing and may alter or deliberately squander one of the richest natural resources in the world to explore a venture with lower value that is less fair, equitable and sustainable. We emphasize that the country has the greatest diversity of freshwater fish in the world, a precious heritage that maintains ecosystem goods and services and is expressed in terms of its economic, cultural, aesthetic, and scientific value. We also emphasize that biodiversity covers phenomena that extends beyond simple species number, as it is related to our understanding of genotypes, phenotypes, organisms, species, interactions, and evolutionary processes that historically occur in a heterogeneous and integrative biosphere. In Brazil and other tropical countries, such in-depth concepts are comparatively poorly understood, particularly for freshwater environments. Therefore, if Brazilian society is concerned with its well being in the long duration and is committed to flourishing in cultural terms, regional biodiversity must be appreciated as a natural capital and preserved (UNITED NATIONS 1992). This approach is most certainly in conflict with management strategies that simplify the starter and propagation of nonnative species or ignore its magnitudes (PELICICE et al., 2014).

4.3. Legal procedures for environmental licensing of aquaculture

According to ALBANEZ et al. (2009), people interest on aquaculture in waters of the Union, guided by the article 3rd of the Decree number 4,895 from 2003 (BRAZIL, 2003), will need to conduce, by means of the State Office of the Federation Unit in which the aquaculture project is centered, petition for the charter of physical environments use to the Special Secretariat of Aquaculture and Fisheries (SSAF/PR), as well as the exclusive project prepared with the activity to be designed according to the Interministerial Normative Instruction number 6 from May 28th 2004 (BRAZIL, 2004).

The authorization to use physical space in water of the Union domain with aquaculture purposes, following what is established on the Decree number 4,895 from 2003, is imperative, and it is not consented to the possessor the installment or lease of the mentioned area. The license of use of aquaculture areas is regulated by the Interministerial Normative Instruction number 6 from May 31st 2004, Chapter III, On Aquaculture Areas. In this sense, checked the methodological adaptation of the project by SSAF/PR, it will be subject to the National Water Agency (ANA) when competing to IBAMA and to the Maritime Authority with competence over the area where the enterprise will be established to proceed with the appreciation and liquidator proclamation number NWA, when required by SSAF/PR, will enunciate the preventive concession for reservation purposes for the hydric availability that allows to the investors the idealization of the required use, according to the Article 6th of the Law number 9,984 from July 17th 2000 (ALBANEZ et al, 2009).

The civil responsibility of those who transgress legal aspects related to the environment is established on the Law number 9,605 from February 12nd 1998 (BRAZIL, 1998). The responsibility is not given only to who transgressed the law, but also to those that knew the criminal conduct of others and did not impede their practice. The responsibility of the legal entity does not exclude the responsibility of the physical entity.

The Decree number 3,179 from September 1999 (BRAZIL, 1999),

which regulates the Law of Environmental Crimes, provides the specifications of the actions applicable to the conduct and activities harmful to the environment. The following agricultural activities are considered crimes against the environment: perishing of aquatic fauna specimens, destruction or damage of permanent preservation forests; pollution of any nature, constructions, renovation, expansion, installation or operation of establishments, works or other potentially polluting services, without license or authorization of the competing environmental agencies.

The expansion of consideration related to the environmental discussions and ascending participation of social collectivity originated modification on the governmental structures that stablishes popular tools of natural resources management, which may be highlighted in Laws, Decree and Ordinances.

• The Federal Constitution of 1988 of Federal Republic of Brazil treats about the goods of Union and States on the Article number 20:

- Of the goods of Union and States

Art. number 20. Belong to the Union:

III – lakes, rivers and any water currents in lands of its domain, or that bathes more than one State that serve the limits with other countries or that extends to foreign territory or that came from them, as well as margin territories and river beaches.

Art. number 21. Compete to the Union:

XIX – institute national system of management of hydric resources and define criterion for granting use rights.

Art. number 22. It is incumbent to the Union legislate on:

IV – waters,

Single paragraph – Complementary law may authorize the States to legislate on specific issues of matters related in this article.

Art. number 23. It is common competence of the Union, Stated, Federal District and municipalities:

VI – protect the environment and combat any pollution;

VII – protect forests, fauna and flora;

VIII – foment the agricultural production and organize the food supply.

Art. number 26. It is included among the goods of State:

I – the surface or groundwater, flowing, emerging or in waters deposits, except in this case as set forth in the law, those arising from works of the Union.

- Of the Environment

Art. number 225. Everyone has the right to an ecologically balanced environment, good of common use and essential to the healthy quality of life, imposing to the Public Power and to the collectivity the duty to defend it for the present and future generations;

• The Decree-Law number 24,643 from July 10th 1934 about the code of Waters – Establishes the possible and different uses of water in general, as well as its prosperity.

• Decree-law number 221 (Code of Fishing, rom February 28th 1967, that revoked the Decree-law number 794, from October 19th 1938), on its:

Chapter IV…

Art. number 34. Prohibits the importation or exportation of any aquatic species in any evolution stage, as well as the introduction of native or exotic species on inland waters without previous authorization of the competing agency, in this case the Brazilian Institute of Environment a Renewable Natural Resources (IBAMA).

Art. number 36. Obligates the decision-making protecting the ichthyofauna by owners and electricity dealers.

Title VI – Of aquaculture and its trading

Art. number 50. The Government must stimulate the creation of federal, state and municipal stations of biology and aquaculture, as well as provide technical assistance to private

Art. number 51. Establishes that the record of fish farmers will be maintained in across the country (Ordinance IBAMA number 95, August 3rd 1993).

Art. number 61 and number 67. Determine the regulation of pisciculture units in relation to production, commercialization, importation and exportation of alive fishes and eggs; of stocking and restocking of aquatic environments and fish farmers record.

Art. 68. Establishes that dams of rivers and streams must have as obligatory complement works that allow the conservation of the river fauna, either facilitating the fish passage or installing pisciculture stations.

• Decree-law number 1,695 from December 13th 1995 and regulated by the Decree number 2,869 from November 9th 1998, which dispose about the authorization of aquaculture exploration on public water belonging to the Union.

• Law number 9,433 from January 8th 1997 that institutes the National Policy of Hydric Resources and creates the National System of Hydric Resources Management. Art. 34 creates the National Council of Hydric Resources.

• Decree Lay number 2,869 from December 9th 1998 regulates the cession of public Waters to the exploration of aquaculture.

• Law number 9,795 from April 27th 1999. Disposes about the environmental education, institutes the National Policy of Environmental Education and provide other measures.

• Decree number 3,057 from May 13th 1999. Creates the Integration Committee of Hydric Infrastructure works and provides other measures.

- Law number 9,984 from July 17th 2000. Disposes about the creation of National Water Agency (ANA), federal entity that actualizes the National Policy of Hydric Resources and coordinates the National System of Hydric Resources Management and provides other measures.

- Decree number 3,945 from September 28th 2001. Creates, in the structure of Environment Ministry, the Department of Genetic Patrimony, which will perform the function of Executive Secretary of Management Council.

- Law number 10,881 from June 9th 2004 – Disposes about management contracts among the National Water Agency and entities that delegate functions of the Water Agency relative to the management of Hydric Resources of the Union domain and provide other measures, Brasília. Law number11,959 from June 29th 2009 – that disposes about the National Policy of Sustainable Development of Aquaculture and Fishing, regulates the fishing activities, repeals the Law number 7,779 from November 23rd 1988 ad devices of the Decree-Law number 221 from February 28th 1967 and provides other measures.

- Resolution number 001 of the National Council of the Environment (CONAMA) from January 23rd 1986 – that regulates the elaboration and presentation of the Environmental Impact Study (EIA) / Environmental Impact Report (RIMA), with direct interest on aquaculture in the Article number 2, items VII and XII.

- Resolution CONAMA number 237 from December 19th 1991 – that establishes norms for the environmental licensing, localization, construction, extension, modification and operation of enterprises and activities that use environmental resources considered effective or potentially polluting as well as enterprises capable to cause environmental degradation anyway.

- Resolution CONAMA number 357 from March 17th 2005 - that disposes about the classification of water bodies and environmental guidelines for its framework, as well as establishes the conditions and standards the effluent discharge, and provides other measures.
- Resolution CONAMA number 413 from June 26th 2009 – that disposes about the environmental licensing of aquaculture, and provides other measures.
- Ordinance of the Brazilian Institute of Environment and Renewable Natural Resources (IBAMA) number 08 from February 2nd 1996 - that establishes general norms for fishing on the Amazon River watershed.
- Ordinance IBAMA number 119 from October 17th 1997 – that establishes norms and procedures for the introduction and reintroduction of fishes, crustaceans, mussels and algae for aquaculture.
- Ordinance of IBAMA number 116 from August 17th 1998 - that establishes norms and procedures for the obtainment of the fish farmer record.
- Ordinance IBAMA number 136 from October 14th 1998 – that establishes the record of fish farmer at IBAMA.
- Ordinance IBAMA number 145-n from October 29th 1998 – that establishes norms for the introduction, reintroduction and transference of fishes, crustaceans, mussels and aquatic macrophytes for aquaculture purposes, excluding ornamental species of animals. Alteration: Ordinance IBAMA number 27 from May 22nd 2003.
- Ordinance IBAMA number 11 from January 30th 2004 – Creates the Technical Work Group to monitor, discuss, evaluate and propose management measures related to piracema in the Paraguay River watershed.
- Ordinance IBAMA number 83 from November 6th from 2006 – Creates the Work Group – GT of Incidental Catch on the Fishing Activity.

- Ordinance IBAMA number 48 from September 2007 – Establishes norms for fishing during the period of protection to the natural reproduction of fishes on the watershed of Amazon River, rivers of Marajó Island and on the watershed of rivers Flexal, Cassiporé, Calçoene, Cunani and Uaçá in the state of Amapá.
- Ordinance IBAMA number3 from January 28th 2008 – Establishes norms for fishing on the Paraguay River watershed in the states of Mato Grosso and Mato Grosso do Sul.
- Ordinance of Ministry of Agriculture in Brazil (MAPA) number 185 from May 13th 1997 – Approves the Technical Regulation of Identity and Quality of Fresh Fish (Entire and Gutted).
- Interministerial Ordinance of Ministry of Fishery (MPA) number 2 from November 13th 2009 – Regulates the System of Shared Management of the sustainable use of fisheries treated by Decree number 6,981 from October 13th 2009.
- Ordinance of Ministry of Labor and Employment (MTE) number 547 from March 11th 2010 – Establishes, in the scope of MTE, the Special Cadastre of Fish Colonies.
- Interministerial MPA and Ministry of Environment (MMA) number 4 from December 2012 – Approves the Internal Regiment of the Technical Committee of the Shared Management of Fisheries.
- Interministerial Ordinance (MPA and MMA) number7 from December 21st 2012 – Creates the Permanent Committee of Fishery Management and Sustainable Use of Resources on the Amazon basin – CPG Amazon Basin in order to advise the Ministries of Fishery and Aquaculture and Environment in the sustainable use of resources of Amazon basin.

- Interministerial Ordinance of Technology (MCTI) and (MPA) number 35 from January 13th 2013 – Institutes the Interministerial Technical Committee of Sciences, Technology and Innovation on Fishery.
- Ordinance ICMBio number 198 from June 19th 2013 – Approves the National Plan of Action for the Conservation of endangered Revalidate fish comprising fifty three threatened species, establishing its general objective, specific objectives, actions, deadlines, modes of execution and supervision.

5. MATERIAL AND METHODS

The work was conducted in the municipalities Ariquemes and Colorado do Oeste, state of Rondonia, West Amazon of Brazil. Twenty fish farmers were interviewed in the first municipality and 15 in the second.

The study had a qualitative and quantitative focus, according to the theory proposed by SAMPIERI et al., (2006). In these focus the sample size is not considered important in a probabilistic perspective because the goal of the research was not to generalize the study results for a broader population.

5.1. Municipality of Ariquemes

It is located at the latitude 09° 54' 48" South and longitude 63° 02' 27" West at 142 m high. The population has approximately 90,354 inhabitants (IBGE/2010). The area is of 4,427 km² and is situated on the North-Central portion of the state, at 203 Km from the capital Porto Velho (Figure 1). The municipality is inserted on the watershed of Jamari River. Ariquemes is an expressive municipality regarding livestock and the production of coffee, cocoa, guarana and cereals. It has the largest roofless mining of the planet. The municipality groups several industries of multiple segments, generating an economy divided for a population that surpasses 90 thousands of inhabitants. The Gross National Product (PIB) of the city is R$ 1,005,152.00 and the PIB per capita is R$11,883.90.

5.2. Municipality of Colorado do Oeste

Colorado do Oeste is a Brazilian municipality located on the state of Rondonia 760 Km far from the capital Porto Velho and is situated on the South of the state (Figure 1). It is located at the latitude 13° 07' 00" South and longitude 60° 32' 30" West at 460 m high. The population estimated by IBGE (2010) was about 23,000 inhabitants. It has an area of 1,442.4 km², representing 0.65% of the state. The climate is hot and humid with mean annual temperature of 23 °C, maximum of 33 °C and minimum of 12°C, with thermal amplitude of 10 °C. The mean annual rainfall is approximately 2,400 mm. The municipality is inserted on the watershed of Guaporé River. The Human Development Index (IDH) of the municipality is of 0.74. The economy is based on the production of soy, fish and cattle. The PIB per capita was 12,158.75 (IBGE, 2010).

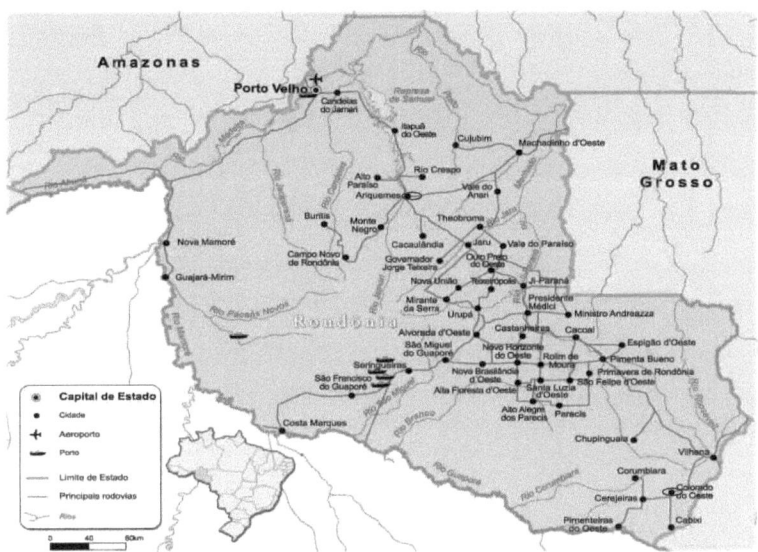

Figure 1 Location of the municipalities Ariquemes and Colorado do Oeste in the state of Rondonia, Brazil. Source: http://www.guiageo.com/rondonia.htm.

5.3. Data collection

The study was grounded in a qualitative and quantitative focus of the scientific research according to the theory proposed by SAMPIERI et al., (2006). For these focus the sample size is not considered important in a probabilistic perspective, because the interest of the research was not to generalize the study results for a broader population. Then, the factors that interfered to determined or suggest the number of cases that compounded the sample were considered, since the qualitative and quantitative study are dynamic and is subject to the way collections were made.

The study was called depth case study considering a study compounded from ten to twenty cases of fish farmers of each studied municipality (Table 2). Samples were considered non-probabilistic and driven.

Samples allowed a deep initial immersion in the field, admitting the obtainment of cases scientifically interesting for the research. The selection of cases did not depend on the probability for reasons related to the research characteristics.

Interviews were made with the aid of questionnaires compounded by qualitative and quantitative questions classified according to knowledge about social, economic and environmental requirements for the fishing activity.

Table 3 exposes the population of fish producers and the sample size used for the calculus of Environmental Management Performance.

Table 3 Size of population and sample of fish farmers from Ariquemes and Colorado do Oeste on 2012.

Municipality	Population	Sample	Sample (%)
Ariquemes	150	20	13.33
Colorado do Oeste	57	15	26.31

Source: research data.

For the collection of data structured interviews were made by means of questionnaires prepared from Table 4.

Table 4 Summary of the questionnaire that evaluated the performance of Environmental Management on pisciculture, Ariquemes, on 2012.

Themes			
1. Property rights and attention to legislation	Yes	No	Na
Does the farm have an operation license? (The environmental licensing for aquaculture on the Federal domain has the IBAMA as competent agency and obeys the pertinent environmental legislation: Resolution CONAMA number 01/1986, Resolution CONAMA number 237/1997 and Resolution CONAMA number 357/2005).			
2. Relations with the community			
Does the farm interfere on the access to areas of public use (leisure and recreation areas that take advantage from the place in function of environmental characteristics)?			
3. Traceability			
Does the farm have records of the fish treatments? (Origin, sanitary state, productivity of the national pisciculture, food			

44

safety and medicines provided to the fish. Represents concurrence and stimulus to quality).

Note: Na = Non-applicable to the question (indecision of the interviewed farmer).

Source: adapted from Carrasco (2006).

The questionnaire used to evaluate the environmental performance on the properties looked for the theoretical approximation to the coefficient of Environmental Management of the farms. The questionnaire was prepared from indications given by the Food and Agriculture Organization of the United Nations (FAO, 1995) for the development of a sustainable pisciculture by means of the Conduct Code for the Responsible Aquaculture (CCAR), according to the theory proposed CARRASCO (2006). The practices of aquaculture management elaborated by the University of Auburn (USA) and used by the Non-Governmental Organization Aquaculture Certification Council (ACC[1]) were considered.

The reliability, coherence and linguistic consistency of the 57 questions corresponded to the degree of coherence with which the credibility characteristics of this questionnaire was evaluated, calculating the total value of Cronbach alpha[2] obtained from the sample (N = 20) of Ariquemes fish farmers. According to DAVIS (1964), questionnaires that measure beliefs, attitudes and socio-demographic values need Cronbach alpha coefficient superior to 0.50.

There were three types of response on the questionnaire: Yes, No, and Indecision of the interviewed people. The color green indicated the acting favorable to the environmental performance, red indicated the unfavorable acting and yellow indicated the non-applicability of the question. The color was independent of the response Yes or Not. In some cases the response Yes was positive and in others was negative. The coefficient of environmental

---

[1] Available in: http://www.aquaculturecertification.org.

45

performance of pisciculture was determined comparing the template with the marked responses. They were analyzed independently for each farm and the degree of environmental performance was determined through the following formula (NORTHEAST BANK IN BRAZIL, 1999): Environmental sustainability of the fish farm = (Green Mark x 100) ÷ (Total of questions – (less) Yellow Marks).

The 57 questions were generated to attend each previous item. These questions were elaborated according to the theory proposed by Carrasco (2006). Some examples of three questions: a) Does the farm have an operation license? (The environmental licensing for aquaculture on the Federal domain has the IBAMA as competent agency and obeys the pertinent environmental legislation: Resolution CONAMA number 01/1986, Resolution CONAMA number 237/1997 and Resolution CONAMA number 357/2005); b) Does the property occupy a Permanent Preservation Area? And c) Does the fish farmer have records that indicate what inputs and treatments each fish lot received?

All questionnaires were digitalized and converted in spreadsheet of Microsoft Office Excel 2007. Then the general estimate for each question made on the evaluated property was elaborated. The measurement of the performance classification of the Environmental Management of the fish farms was presented as percentage of the environmentally positive responses of all properties and of the categorization of the Environmental Management Performance (Table 3).

All responses of each fish farm were inserted on spreadsheets and analyzed with the aid of the software Statistical Package for the Social Science (SPSS), version 11.5 (SPSS 11.5). The responses of all farms were summed to interpret results. The percentages of environmentally positive responses (green color) were obtained ignoring the yellow responses.

46

Results obtained were interpreted considering percentage values of Environmental Management Performance (Table 5). These criteria of classification were established from the theory proposed by Northeast Bank in Brazil (1999) and CARRASCO (2006).

Table 5 Criteria for the classification of Environmental Management Performance on the aquaculture of Ariquemes and Colorado do Oeste on 2012.

Criteria	Rating
Below to 30%	Critical
Between 30.1 and 50%	Bad
Between 50.1 and 70%	Adequate
Between 70.1 and 90%	Good
Higher to 90.1%	Excelent

Source: adapted from Northeast Bank in Brazil (1999) and Carrasco (2006).

6. RESEARCH FUNDING

The research was funded by National Council of Scientific and Technological Development (CNPq Brazil), (Table 6).

Table 6 Period, modalities and total values funded by CNPq.

Periods	Modality	Value (US$)
11/01/2011 to 10/31/2012	Scholarship and additional finantial support	19,649.00
11/01/2012 to 31/10/2013	Scholarship and additional finantial support	17,684.00
Total funded value		37,333.00

Source: research data

7. RESULTS AND DISCUSSION

Aquaculture in the municipalities of Ariquemes and Colorado do Oeste presented two systems of fish production: extensive and semi-extensive.

According to BRAZIL (2008), in relation to the System of Breeding and water use evaluated according to the accumulated water slide, the aquaculture will be classified in: a) Extensive system: practiced in dams and lakes in which there is no control of depth and flow, with production on until one ton per hectare; b) Semi-extensive: fishponds dams with controlled depth and flow of reservoirs, with production of until 6 (six) tons per hectare, and fishponds dams of model recommended by Secretary of Agriculture, Production and Economical and Social Development of the state of Rondonia (SEAPES) (without water renovation); c) Intensive: practiced on fishponds excavated in natural terrain, with production of 6 (six) to 15 (fifty) tons per hectare; and IV – Super Intensive: Cages and Race-ways (concrete tanks of high water flow), with production superior to 15 (fifteen) tons per hectare.

According to BRAZIL (2008), the following definitions and provisions for the aquaculture activities: I – cultivation and farming aquatic organisms, including fishes, mussels, crustaceans, turtles, retiles and aquatic plants through the man intervention in order to increase the production in operations like reproduction, storage, feeding, protection against predators and other; II – Pisciculture: activity of farming fries or fishes on natural and artificial environments with economic, social or scientific purposes; III – Fish farmer: physical or legal entity that dedicates professionally to farm fries or fishes on natural and artificial environments with economic, social or scientific purposes, working independently or linked to associations or cooperatives; IV – Fry farmer: fish farmer who dedicates to the reproduction, hatchery, farming and commercialization of fries; V – breeder or matrix: adult fish, able to procreate, used by the fish farmer to obtain descendants; VI – Reservoir: natural or artificial of surface water such as rivers, lagoons, ponds, channels and others; VII – Dam: water deposit formed through the construction of barriers on natural geographic landforms and/or caused by anthropic action

48

through which rainwater, rivers, streams, with the aim to use them as hydric resource; VIII – fishpond or tank: structure designed and built for aquaculture, excavated or not, coated or not, and with controlled water input and output; IX – Aquaculture area: physical continuous space in aquatic environment, delimited and destined to aquaculture projects, individual or collective; X – Aquaculture park: physical continuous space in aquatic environment, delimited, that comprises a set of aquaculture related areas in which intermediate physical spaces activities compatible with aquaculture may be developed; XI – Cages: equipment for cultivation used within the water mass of a river, lake or reservoir, aquaculture park, constructed and managed according the technical norms of engineering; XII – Native species: species of origin and natural occurrence on Brazilian waters; XIII – Exotic species: species of origin and natural occurrence only recorded for waters in other countries; XIV – Stablished species: allochthonous species that already constitutes and isolate population and reproduces, appearing on scientific or extractive fishing; XV – Hybrid fish: fish obtained from the crossing between species; XVI – Allochtonous: non-originated on the watershed; XVII – Autochthonous species: originated on the watershed; XVIII – Fish stocking: process of introduction of fries or adult fishes on natural or artificial aquatic environments with the aim to populate or repopulate the local water body; XIX – Fish removal: process of removal of fishes and aquatic species cultivated with economic, social and scientific purposes; XX – Spring: place in which the groundwater emerges naturally, even intermittently.

In the extensive system of the studied area commercial feed was rarely used and fishes were fed with agricultural by products or animal wastes. The productivity did not surpass the range from 1,000 to 3,000 Kg of fish/ha/year. The extensive system was characterized by the use of the species Tambaqui

(*Colossoma macropomum*), Tilapia (*Oreochromis* spp.), Pirarucu (*Arapaima gigas*) and Pintado (*Pseudoplatystoma sp.*).

Fish farmers used extruded feed with four different coefficients of guarantee (Table 7). The size of feed pellets for the guarantee coefficient 1 was 08-10 mm, for the coefficient 2 was 15-20 mm, for the coefficient 3 was 10-15 mm and for the coefficient 4 was 04-06 mm. All the three feeds with guarantee coefficient of 28% of crude protein (PB) were used for omnivorous fishes in the fattening stage. The feed with 32% of PB was used for omnivorous on the young phase.

Table 7 Feeds used by fish farmers of Colorado do Oeste city 2012.

	Guarantee coefficients (%)			
	1	2	3	4
PB	28.0	28.0	28.0	32.0
Ether extract	4.0	3.4	4.0	6.8
Crude fiber	4.5	5.7	4.5	4.8
Phosphorus	1.0	1.0	1.0	1.2
Mineral matter	7.5	7.0	7.5	8.0
Calcium	1.7	1.7	1.7	2.1
Moisture	9.2	12.0	9.2	9.2

Source: research data.

The semi-intensive system was characterized by the species Tambaqui, Tilápia and Pirarucu. The creation of Tambaqui in the municipality of Colorado do Oeste presented satisfactory results. The technological package used on the semi-intensive production of Tambaqui was the fishpond with dam in which the production was divided on two phases: growing, with cycle of 60 days and fattening, with cycle of 240 to 300 days. The productivity did not surpass the range of 4,500 to 6,000 Kg of fish/ha/year.

The most frequent sanitary problems were related to diseases caused by fungi and bacteria.

The Table 8 describes the price paid by the fridge by Kg of fish produced on Colorado do Oeste, the amount of fish sold directly to the consumer, the amount of fish sold to the industry, the gross earnings

50

obtained, the net earnings obtained, the tax collected for the state of Rondonia and the tax paid by the industry.

Table 8 Economic data for the aquaculture of Colorado do Oeste city 2010.

Paid by the fridge (US$/Kg fish)	Sold fish (Kg)	Sold to the industry (Kg)	Gross earnings (US$)	Net earnings (US$)	Tax paid by the industry (US$)
1.15	15,000	172,100	223, 603.51	115, 954.00	11, 942.39

Source: research data.

The approximation to the coefficients of Environmental Management Performance obtained through the evaluation of properties of the sample (N=15) ranged from 1.8 to 44.1% (Table 9).

Table 9 Approximation to the coefficients of Environmental Management Performance of properties analyzed on the sample from Colorado do Oeste city 2012.

Number of farm	Environmental Management
01	1.8
02	6.3
03	6.3
04	11.4
05	11.8
06	17.9
07	19.8
08	19.9
09	23.4
10	30.9
11	32.5
12	33.0
13	37.0
14	38.5
15	44.1

Source: data from the current research using the software SPSS 11.5

Data show that nine of the first properties evaluated on Colorado do Oeste presented critical development of Environmental Management, inferior to 30%. The other six properties presented bad performance, between 30 and 50.0% (Table 8).

The approximation to the coefficient of Environmental Management Performance obtained on the evaluation of the properties of the sample (N = 20) ranged from 5.3 to 67.7% (Figure 2).

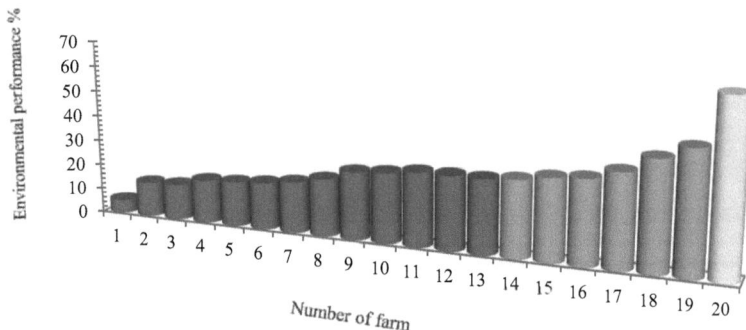

Figure 2 Approximation to the coefficients of Environmental Management Performance of the properties analyzed on the sample of Ariquemes city on 2012. Source: research data using the software SPSS 11.5

Data presented on Figure 3 showed that thirteen farms presented critical Environmental Management Performance, six presented bad performance (between 30 and 50.0%) and one property presented adequate performance (Between 50.1 and 70.0%).

The evaluations of environmental sustainability and the Environmental Management performance are based on legal issues. The development of a suitable administrative and legal framework in order to guarantee the introduction and application of environmentally responsible practices of aquaculture represented by meeting the legal obligations, relation with the community, worker safety, conservation of protected areas, management of the fishpond effluents, managements of sediments of fattening and nursery

52

tanks, management of medicines and treatments received by each fish lot (Table 10).

In relation to the worker safety and relationship with the, 100.0% of the interviewed fish farmers answered that they did not provide trainings about general security, personal hygiene and first aid to the employees on the fishing activity of their properties (Table 10).

These difficulties may be related to inadequacies of the management tools faced with the fish farmers reality: undercapitalization, because they cannot have access and take advantages of the modern technologies of information; low coefficient of formal education of fish farmers; lack of culture that creates environment propitious to the adoption of new technologies of management and, finally, lack of adequate training of the technicians responsible by the assistance to the producers.

In relation to the conservation of protected areas, 60.0% of the interviewed fish farmers answered that they occupy permanent preservation areas, which are represented by riparian vegetation, hill tops and floodplains (Table 10). These results may be related to the fact that the current legislation for the preservation of permanent areas has not a guaranteed application, even it had been reformulated. Its interpretation is hampered by the lack of scientific knowledge of the fish farmers and also by the influence of local mores.

For the group of questions that analyzed data related to the management of effluents of fishponds, 100.0% of the fish farmers did not dispose sediments adequately to avoid the erosion of ponds for fish fattening. 66.7% of the interviewed fish farmers did not remove the sediments of fishponds, weir and other areas within the property (Table 10).

These results may be related to the fact that the Resolution CONAMA number 357 from 2005, which disposes about the classification and environmental guidelines for adequacy of water bodies, as well as establishes conditions and standards for effluents discharge, does not stipulates maximum concentrations of solids (total, fixed and volatile) to verify the quality of effluents released on water bodies. After the fish collection, the farmers must the 30% of the water decant and then eliminate it on the water bodies.

In relation to the group of questions that analyzed data regarding the management of medicines and chemicals prohibited on Brazil and other countries, 93.3% of the interviewed fish farmers answered they do not know the list of medicines and chemical prohibited in Brazil and in the world (Table 10).

These results may be related to the fact that in Brazil there are no statistics in respect to the amount of antibiotics commercialized for the animal production. These difficulties may be related to the inadequacy of management tools used by fish farmers such as the access to modern technologies of information; low coefficient of formal education of fish farmers and lack of training adequate to technicians responsible by the assistance to producers.

More than 93.0% of interviewed fish farmers on the sample said that records that allowed them to know what inputs were used on the fish cultivation and which treatments each fish lot receives (Table 10).

The results obtained may be related to the fact that the lack of organization of aquaculture is the main obstacle for its development, because the lack of management to mitigate the environmental, social and economic barriers have departed the sector from the coefficients of environment management performance superior to 50%. The absence of Traceability System prevents that the aquaculture of Colorado do Oeste assumes other great objective: to gather producers and companies for an organized and participatory management.

In 100.0% of the studied cases the legal obligations were attended by fish farmers (Table 10). They had affirmed that obey to what is established by the environmental legislation relevant to the environmental licensing for fishery according to legal assumptions established on the Resolution CONAMA (National Council of the Environment) number 01 from 1986 and on Resolution CONAMA number 237 from 1997. In the State Plan the responsibility by the environmental licensing is passed to the Environmental State Agencies, obeying the current state legislation that cannot be permissible than what is established on the Federal Law that regulates the environmental licensing.

The relationship with the community faced problems because 100.0% of the interviewed fish producers answered that their property does not contribute to the welfare of the community and socio-environmental (health, recreation and education).

These results may be related to the fact that great part of the fish production on Colorado do Oeste city comes from family producers and a small portion comes from the business aquaculture. The lack of technological and management training these two processes is still fragile. There is also a lack of public participation. Mechanisms of governance must not allow that political interest exceeds technical-scientific and sociocultural issues. This leads to a latent conflict between the need to protect the environment (on the limits required by the forest legislation), and the need of rural production required by the social function of the properties.

Table 10 Significant answers for groups of questions, Colorado do Oeste city 2012.

Question	% of yes	% of no
Compliance with legal obligations	100	0.0
Relationship with the community	0.0	100
Worker safety	0.0	100
Conservation fo protected areas	60.0	40.0
Management of the fishpond effluents	0.0	100
Management of the fishpond sediments	33.3	66.7
Management of prohibited medicines and chemists	6.7	93.3
Treatment received by each fish lot	6.7	93.3

Source: research data.

From the 20 interviewed fish farmers from Ariquemes city, 13 affirmed (65%) that do not have environmental license for the management of pisciculture on their farms. Seven fish farmers (35%) answered they have environmental license (Table 11). The confidence interval was of 95% of probability.

Table 11 Conditions of the environmental license of pisciculture operation on the farms.

Valid	Frequency	%
No	13	65.0
Yes	07	35.0
Total	20	100.0

Source: research data using SPSS 11.5

Of the 20 fish farmers interviewed in the city of Porto Velho, 16 said (80%) did not recycle water fish farming on his farm. Four fish farmers (20%) said they recycle water for the management of fish farming on his farm (Table 12). The confidence interval was 95% probability.

Table 12 Frequency of answers for the use of water flowing from the fishponds.

Valid	Frequency	%
No	16	80.0
Yes	04	20.0
Total	20	100.0

Source: research data using SPSS 11.5

From the 20 interviewed fish farmers on Ariquemes, 15 affirmed (75%) that they use fish species from other regions and countries on the pisciculture of their farms. Five fish farmers (25%) answered they do not use species from other regions (Table 13). The confidence interval was of 95% of probability.

Table 13 Frequency of answers regarding the use of fish from other regions.

Valid	Frequency	%
No	15	75.0
Yes	05	25.0
Total	20	100.0

Source: research data using SPSS 11.5

According to the Resolution CONAMA number 413/2009 the operating companies that does not have environmental license must regulate their situation according to the agency of environmental licensing. The situation regularization will be made through the obtainment of the Operation License – LO, single license (by simplified procedure) or only a record, in cases in which is possible to dispense the license. For cases in which there is the emission of LO or Installation and Operation License (Lio), farmers will be obligated to provide the documents containing at least: a general vision of the relevant environmental studies, mitigation and measures for environmental protection according to the criteria of the agency of environmental licensing and tools of management to guarantee the execution of such measures.

Results revealed an alpha Cronbach coefficient of 0.831 for the total of 57 questions of this questionnaire adapted and validated on sample of 20 fish farmers of Ariquemes and also used to make the interviews with fish farmers from Colorado do Oeste. This index revealed that the interviews were reliable, once the used questionnaire proved to be reliable, consistent and linguistically concise.

8. CONCLUSION

. The percentage coefficients of environmental performance showed that the aquaculture on the state of Rondonia is environmentally unsustainable. This means that only the ordered action of fish farmers, industries and institutions of education, research and extension may determine instructions and practices of an environmentally responsible management.

. 93.3% of the interviewed fish farmers answered they do not know the list of medicines and chemists prohibited for the pisciculture in Brazil.

. Approximately 80% of the fish farmers answered they do not recycle the water flowed from tanks of fish.

. Approximately 75% of the fish farmers answered they use species of fish from other regions and countries.

. These percentage data may be related to the inadequate training of fish farmers.

. The value 0.831 of the alpha Cronbach coefficient obtained on the research indicated reliability, consistence and linguistic coherence of the questionnaire used for the collection of data.

. There are no processes for the identification of areas propitious to the pisciculture in Rondonia trough an inter-institutional and participatory methodology with the support of Local and State Committees that respect the interests of local communities.

. There is no territorial planning of the continental Aquaculture in Brazil.

. Technological principles are lacking in Rondonia, which may guarantee the quality of the Environmental Management of aquaculture and also the substantiation of a conduct code for the environmentally responsible fishing according to technical assumptions instituted by FAO.

9. REFERENCES

ALBANEZ, João R.; ALBANEZ, Ana C. M. P; PEREIRA, Maria C. **Environmental legislation applied to the pisciculture**. Lavras, 2009, 94p.

ANDRADE, Daniel Caixeta; ROMEIRO, Ademar Ribeiro. Environmental Degradation and Economic Theory: Some Reflections on the "Economy of Ecosystems". **Economy**, v. 12, n. 1, 2011.
AZEVEDO-SANTOS, V.M., RIGOLIN-SÁ, O. & PELICICE, F.M. Growing, losing or introducing? Cage aquaculture as a vector for the introduction of nonnative fish in Furnas Reservoir, Minas Gerais, Brazil. Neotrop. **Ichthyol.**, 9, 915-919, 2011.

BRAZIL. Federal Decree number 23,672 from January 2nd 1934. Approving the hunting and fishing code. **Official Gazette of the Union,** Rio de Janeiro, January   2nd   1934.   Available   in: < http://www2.camara.leg.br/legin/fed/decret/1930-1939/decreto-23672-2-janeiro-1934-498613-publicacaooriginal-1-pe.html>. Access in: december 13th 2013.

BRAZIL. Federal Decree number 23,793 from January 23rd 1934. Decreeing the forest code. Brasília, DF, 1934a. **Official Gazette of the Union.** Available in:<http://www.planalto.gov.br/ccivil_03/decreto/1930-1949/d23793.htm>. Access in: december 20th 2011.

BRAZIL. Decree number 28,840 from November 8th 1950. Declares the national underwater shelf integrated to the national territory and provides other measures. **Official Gazette of the Union,** Rio de Janeiro, November 8th 1950. Available in:<http://www2.camara.leg.br/legin/fed/decret/1950-1959/decreto-28840-8-novembro-1950-329258-publicacaooriginal-1-pe.html>. Access in: december 14th 2013.

BRAZIL. Decree-law number 221 from December 28th, 1967. Disposes about the protection and incentives to the fishing and provide other measures.

Brasília, DF, 1967. **Official Gazette of the Union.** Available in:<http://http://www.planalto.gov.br/ccivil_03/decreto-lei/del0221compilado.htm>. Access in: february 5th 2012.

BRAZIL. Decree-law number 44 from November 18th,1966. Alters the limits of territorial water of Brazil, establishes a contiguous zone and provide other measures. **Official Gazette of the Union,** Brasília, November 18th 1966. Available in:<http://www2.camara.leg.br/legin/fed/declei/1960-1969/decreto-lei-44-18-novembro-1966-378095-publicacaooriginal-1-pe.html>. Access in: december 14th 2013.

BRAZIL. Decree number 62,837 from June 6th, 1968. Disposes about the exploration and research on the underwater shelf and internal waters and provide other measures. **Official Gazette of the Union,** Brasília, June 6th 1968. Available in:<http://www2.camara.leg.br/legin/fed/decret/1960-1969/decreto-62837-6-junho-1968 404418-publicacaooriginal-1-pe.html>. Access in: december 14th, 2013.

BRAZIL. Decree-law number 553 from April 25th, 1969. Alters the limits of Brazilian territorial waters and provide other measures. **Official Gazette of the Union,** Brasília, April 25th 1969. Available in:< http://www2.camara.leg.br/legin/fed/declei/1960-1969/decreto-lei-553-25-abril-1969-376473-publicacaooriginal-1-pe.html>. Access in: december 14th 2013.

BRAZIL. Decree-law number 1,098 from March 25th 1970. Alters the limits of territorial waters of Brazil and provide other measures. **Official Gazette of the Union,** Brasília, April 25th, 1970. Available in: <http://www.planalto.gov.br/ccivil_03/decreto-lei/1965-1988/del1098.htm>. Access in: december 14th 2013.

BRAZIL. Law number 6,938 from August 31st, 1981. Establishes tools that ensure the rights to the balanced environment such as evaluation of the environmental impacts, the environmental licensing and the review of polluting activities, the zoning and the environmental supervision. **Official Gazette of the Union,** Brasília, September 2nd 1981. Available in:< http://www.planalto.gov.br>. Access in: december 5th, 2011.

BRAZIL. Constitution (1988). **Constitution of the Federative Republic of Brazil.** Brasilia: Senate, 1988. 168p.

BRAZIL. Law number 8,171 from January 17th 1991. Disposes about the agricultural policy. **Official Gazette of the Union**, Brasília, January 17th 1991. Available in: <http://www.planalto.gov.br/ccivil_03/leis/l8171.htm>. Access in: november 1st 2013.

BRAZIL. Law number 8,617 from January 4th, 1993. Disposes about the territorial waters, contiguous zone, exclusive economic zone and the Brazilian continental shelf and provide other measures. **Official Gazette of the Union**, Brasília, January 4th 1993. Available in: <http://www.planalto.gov.br/ccivil_03/leis/l8617.htm#art16>. Access in: december 14th, 2013.

BRAZIL. Decree-law number 1,695, of December 13th, 1995. Regulates the fish-farming in public waters belonging to the Union and other measures. Official Gazette of the Union. Available in: <http://presrepublica.jusbrazil.com.br/legislacao/112455/decreto-1695-95>. Access in: december 14th, 2013.

BRAZIL. Law number 9,433 of January 8th, 1997. It establishes the National Water Resources Policy, Creates the National System of Water Resources Management and other Measures. **Official Gazette of the Union**, Brasilia, January 8th, 1997. Available at: <http://www.planalto.gov.br/ccivil_03/leis/l9433.htm>. Access in: november 21 th, 2012.

BRAZIL. Law number 9,605, of February 12th, 1998. Provides for penal and administrative sanctions derived from conduct and activities harmful to the environment, and other provisions. **Official Gazette of the Union.** Available at:<http://www.planalto.gov.br/ccivil_03/leis/l9605.htm>. Access in: november 28th, 2014.

BRAZIL. Decree Law number 2,869 of december 9th 1998. It regulates the transfer of public waters for exploration aquaculture, and other measures. **Official Gazette of the Union.** Available in: <http://www.planalto.gov.br/ccivil_03/decreto/d2869.htm>. Access in: november 22th, 2014.

BRAZIL. Decree number 3,179 of September 21nd 1999. Regulates Law of Environmental Crimes and has the specifications of the actions applicable to conduct and activities harmful to the environment. **Official Gazette of the Union**, Brasilia, September 22, 1999. Available at:

<http://www2.camara.leg.br/legin/fed/decret/1999/decreto-3179-21-setembro-1999-344968-normaatualizada-pe.pdf>. Access in: november 22th, 2014.

BRAZIL. Law number 9,795 of 27 april 27nd 1999. It has about environmental education, establishing the National Environmental Education Policy and other measures. **Official Gazette of the Union.** Available in:<http://www.planalto.gov.br/ccivil_03/leis/l9795.htm>. Access in: november 22th, 2014.

BRAZIL. Decree number 3,057 of may 13th 1999. Creates the Integration Committee of Water Infrastructure works, and other measures. **Official Gazette of the Union.** Available in:< http://www.planalto.gov.br/ccivil_03/decreto/d3057.htm>. Access in: november 22th, 2014.

BRAZIL. Law number 9,984 of July 17th 2000. It provides for the creation of the National Water Agency, federal entity implementation of the National Water Resources and coordination of the National Water Resources Management System Policy, and other measures. **Official Gazette of the Union,** Brasilia, July 17, 2000. Available in:<http://www.planalto.gov.br/ccivil_03/leis/l9984.htm>. Access in: november 26th, 2014.

BRAZIL. Law number 9,985 of July 18th 2000. Regulates article. 225, § 1, items I, II, III and VII of the Constitution, establishing the National System of Protected Areas of Nature and other measures. **Official Gazette of the Union,** Brasilia, 18th July 2000. Available in:< http://www.planalto.gov.br/ccivil_03/leis/l9985.htm>. Access in: november 26th, 2014.

BRAZIL. Law number 10, 881 of 09 June 2004. Provides for the management contracts between the National Water Agency and delegators entities of Water Agencies of functions related to water resources management of the union domain and other measures. **Official Gazette of the Union,** Available in:<http://www.planalto.gov.br/ccivil_03/_ato2004-2006/2004/lei/l10.881.htm>. Access in: november 27th, 2014.

BRAZIL. Decree number 3,945 of september 28th 2001. Establish, to the structure of the Ministry of Environment, the Department of Genetic Heritage, which will hold the Executive Secretariat function of the Board of Management. **Official Gazette of the Union.** Available at:

<http://www.planalto.gov.br/ccivil_03/decreto/2001/d3945.htm>. Access in: november 27th, 2014.

BRAZIL. Decree number 4,895 from november 25th 2003. Disposes about the authorization for the use of physical spaces of water bodies belonging to the Union domain with aquiculture purposes, and provide other measures. **Official Gazette of the Union**, Brasília, November 26th 2003. Available in:< http://www.planalto.gov.br/ccivil_03/decreto/2003/d4895.htm>. Access in: december 15th 2013.

BRAZIL. Interministerial Normative Instruction number 6 from may 28th 2004. Establishes the complementary norms for the authorization of the use of physical spaces of water bodies belonging to the Union domain with aquiculture purposes, and provide other measures. **Official Gazette of the Union**, Brasília, May 31st 2004. Available in:<http://www.jusbrazil.com.br/diarios/598900/pg-6-secao-1-diario-oficial-da-uniao-dou-de-31-05-2004>. Access in: august 8th, 2014.

BRAZIL. Interministerial Normative Instruction number 7 from april 28th 2005. Establish guidelines for the implantation of parks and areas of aquiculture due to the Article 19 of Decree n.4895 from November 25th 2003. **Official Gazette of the Union**, Brasília, april 28th 2005. Available in: <http://www2.camara.leg.br/legin/marg/instni/2005/instrucaonormativaintermi nisterial-7-28-abril-2005-536935-norma-mma_seape.html>. Access in: december 15th, 2013.

BRAZIL. Law n. 1,861 from January 10th 2008. Dispose, define and discipline the Pisciculture on the state of Rondonia and provides other measures. **Official Gazette of the Union**, Porto Velho, january 10th 2008. Available in: http://www.sedam.ro.gov.br/images/stories/psicultura/lei_de_piscicultura.pdf> . Access in: december 14th, 2013.

BRAZIL. Law number 11,959 from june 29th 2009. Dispose about the National Policy of Sustainable Development of Aquaculture and Fishing, regulates the fishing activities, revokes the Law n. 7779 from November 23rd 1998 and devices of the Decree-Law n. 221 from February 28th 1967, and provides other measures. **Official Gazette of the Union.** Available in:<http://www.planalto.gov.br/ccivil_03/_ato2007-2010/2009/lei/l11959.htm>. Access in: december 14th, 2014.

BRAZIL. Ministry of Fishery and Aquaculture. **Statistical report of fishery and aquaculture:** Brazil. Brasília: MPA, 2010. 128p. Available in:<http://www.mpa.gov.br/mpa/seap/jonathan/mpa3/dados/2010/docs.Acces s in: march 28th, 2011.

BRITTON, J. Robert; ORSI, Mário Luís. Non-native fish in aquaculture and sport fishing in Brazil: economic benefits versus risks to fish diversity in the upper River Paraná Basin. **Reviews in Fish Biology and Fisheries**, v. 22, n. 3, p. 555-565, 2012.

BORGES, Aurélio. F. et al. Environmental development of pisciculture on Brazilian Western Amazon. **Global Science and Technology**, v. 6, n. 1, 2013a.

BORGES, Aurélio F. et al. Licensing and environmental performance of aquaculture in Western Amazon. **Global Science and Technology**, v. 6, n. 3, 2013b.

BUENO, Guilherme W. et al. "Implementation of aquaculture parks in Federal Government waters in Brazil." **Reviews in Aquaculture**, 2013, v. 5, 1-12.

CARRASCO, Sandra Clemencia Pardo. **Diagnosis of the environmental state and elaboration of a model for environmental management of the pisciculture on Castilla la Nueva, meta Colombia.** 2006. (Doctoral thesis). Department of Production Engineering: Federal University of Santa Catarina, Florianópolis, Brazil, 2006, 160p.

CASAL, C.M.V. Global documentation of fish introductions: the growing crisis and recommendations for action. **Biol. Invas.**, 8, 3-1, 2006.

CLARK, John R.; GARCIA, S. M.; CADDY, J. F. **Integrated management of coastal zones**. Rome: Food and Agriculture Organization of the United Nations, 1992.

CONAMA. National Environmental Council. **Resolution number 01** from January 23rd 1986. Dispose about basic criteria and general guidelines for the evaluation of environmental impacts. Brasília, DF, 1986. Available in: <http://www.mma.gov.br/port/conama/legiabre.cfm?codlegi=23>. Access in: july 20th 2012.

CONAMA. National Environmental Council. **Resolution number 237** from December 19th 1997. Dispose about the review and complementation of procedures and criteria used for the environmental licensing. Brasília, DF, 1997. Available in:< http://www.mma.gov.br/port/conama/legiabre.cfm?codlegi=237>. Access in: july 21st 2012.

CONAMA. National Environmental Council. **Resolution number 357** from March 17th 2005. Dispose about the classification of water bodies and environmental guidelines, besides providing other measures. Brasília, DF, 2005. Available in: < http://www.mma.gov.br/port/conama/legiabre.cfm?codlegi=459>. Access in: july 22nd 2012.

CONAMA. National Environmental Council. **Resolution number 413** from June 26th 2009. Dispose about the environmental licensing of aquaculture and provide other measures. Brasília, DF, 2009. Available in: <http://www.mma.gov.br/port/conama/legiabre.cfm?codlegi=608>. Access in: february 19th 2009.

CONFERÊNCIA DAS NAÇÕES UNIDAS SOBRE MEIO AMBIENTE E DESENVOLVIMENTO. **Agenda 21**. Brasília: Senado Federal, 2001.

CRONBACH, Lee J. Coefficient alpha and the internal structure of tests. **Psychometrika**, v. 16, n. 3, p. 297-334, 1951. Available in:<http:// www.springerlink.com/content/n435u12541475367/>. Access in: december 4th 2011. doi: 10.1007/bf02310555.

DA SILVA, Alexandre Pereira. The new Brazilian claim at sea: the extended continental shelf and the Blue Amazon Project. **Brazilian Journal of International Policy,** v. 56, n. 1, p. 104-121, 2013. Available in:< http://www.scielo.br/pdf/rbpi/v56n1/06.pdf>. Access in: december 13th 2013.

DAVIS, F. B. **Educational measurements and their interpretation**. Belmont: Wadsworth, 1964.

FAO, 1995. **Code of conduct for the responsible fishing**. Roma, 1995. 48p.

____, 1999a. **The state of world fisheries and aquaculture.** Roma. 1998. 254p.

____,1999b. **Guidelines for the responsable fishing**. v. 5. Roma, 1999. 54p.

FLORIANO, Eduardo Pagel. **Environmental management policies**. 3.ed. Editor: Federal University of Santa Maria. 2007. 111p. Available in:< http://www.ufsm.br/dcfl/seriestecnicas/serie7.pdf>. Access in: february 25nd 2011.

IBAMA. Brazilian Institute of Environment and Renewable Natural Resources. **Ordinance number 08 from february 2nd 1996**. Establish general norms for the fishery on the Amazon River watershed. Available in: <http://www.sema.mt.gov.br/index.php?option=com_content&view=article&id =414&itemid=346>. Access in: november 22nd 2014.

IBAMA. **Ordinance number 119 from October 17th 1997**. Establish norms and procedures for the introduction and reintroduction of fishes, crustaceans, mussels and algae with aquaculture purposes. Available in: <http://www.sema.mt.gov.br/index.php?option=com_content&view=article&id =414&itemid=346>. Access in: november 22nd 2014.

IBAMA. **Ordinance number 116 from October 17th 1998**. Establish norms and procedures for the obtainment of the fish farmer record. Available in: <http://www.sema.mt.gov.br/index.php?option=com_content&view=article&id =414&itemid=346>. Access in: november 22nd 2014.

IBAMA. **Ordinance number 145-n from october 29th 1998**. Establishes norms for the introduction, reintroduction and transference of fishes, crustaceans, mussels and aquatic macrophytes for aquiculture purposes, excluding ornamental species of animals. Alteration: Ordinance IBAMA n. 27 from May 22nd 2003. Available in: <http://www.sema.mt.gov.br/index.php?option=com_content&view=article&id =414&itemid=346. Access in: november 22nd 2014.

IBAMA. **Ordinance number 11 from january 30th 2004**. Creates the Technical Working Group to monitor, discuss, evaluate and propose ordering measures referring to piracema on the Paraguay River watershed. Available in: <http://www.sema.mt.gov.br/index.php?option=com_content&view=article&id =414&itemid=346.Availablein:<http://www.sema.mt.gov.br/index.php?option=

com_content&view=article&id=414&itemid=346>. Access in: november 22nd 2014.

IBAMA. **Ordinance number 83 from november 6th 2006**. Create the Working Group-GT of Incidental Captures on the Fishing Activity. Available in: <http://www.sema.mt.gov.br/index.php?option=com_content&view=article&id =414&itemid=346>. Access in: november 22nd 2014.

IBAMA. **Ordinance number 48 from september 25th 2007**. Establish fishing norms for the period of protection to the natural reproduction of fishes on the watershed of Amazon river, rivers of Marajó island and on the watershed of the rivers Flexal, Cassiporé, Calçoene, Cunani and Uaça on the state of Amapá. Available in:< http://www.sema.mt.gov.br/index.php?option=com_content&view=article&id= 414&itemid=346>. Access in: november 22nd 2014.

IBAMA. **Ordinance number 3 from january 28th 2008**. Establish norms for fishing on the Paraguay river watershed on the states of Mato Grosso and Mato Grosso do Sul. Available in: <http://www.sema.mt.gov.br/index.php?option=com_content&view=article&id =414&itemid=346>. Access in: november 22th 2014.

IBGE. **Demographic census**. Brasília, 2010.

ICMBIO. Institute Chico Mendes of Biodiversity Conservation. **Ordinance number 198 from june 19th 2013**. Approves the National Action Plan for the Conservation of killifishes threatened of extinction, establishing the general objective, specific objectives, actions, time for execution, coverage and forms of execution and supervision. Available in: <http://www.sema.mt.gov.br/index.php?option=com_content&view=article&id =414&itemid=346>. Access in: november 22th 2014.

MAPA. Ministry of Agriculture, Livestock and Supply. **Ordinance number 185 from may 13th 1997**. Approves the Technical Regulation of identity and Quality of Fresh Fish (Entire and Gutted). Available in:< http://www.sema.mt.gov.br/index.php?option=com_content&view=article&id= 414&itemid=346>. Access in: november 22nd 2014.

MICROSOFT. **Software Excel 2007**. Available in: <http://office.microsoft.com/en-us/excel-help/up-to-speed-with-excel 007rz010062103.aspx>. Access in: april 11th 2011.

MCTI and MPA. Ministry of Science, Technology and Innovation and Ministry of Fishery and Aquaculture. **Interministerial ordinance number 35 from January 16th 2013**. Institute the Interministerial Technical Committee of Science, Technology and Innovation on Fishery and Aquaculture – CTPA with the aim to establish technical and scientific cooperation. Available in:<http://www.sema.mt.gov.br/index.php?option=com_content&view=article&id=414&itemid=346>. Access in: november 22nd 2014.

MPA and MMA. Ministry of Fishery and Aquaculture and Ministry of the Environment. Interministerial Ordinance number 2 from november 13th 2009. **Regulates the System of Shared management os the sustainable use of fishing resources treated by the Decree number 6,981 from october 13th 2009**. Available in:< http://www.sema.mt.gov.br/index.php?option=com_content&view=article&id=414&itemid=346>. Access in: november 22nd 2014.

MPA and MMA. **Interministerial Ordinance number 4 from december 11th 2012**. Approve the Internal Regiment of the Technical Committee of the Shared Management of Fishing Resources. Available in:< http://www.sema.mt.gov.br/index.php?option=com_content&view=article&id=414&itemid=346>. Access in: november 22th 2014.

MPA and MMA. **Interministerial Ordinance number 7th from december 21st 2012**. Create the Permanent Committee of Fishing Management and Sustainable Use of Resources on the Amazon Basin – CPG Amazon Basin in order to advise the Ministries of Fishery and Aquaculture and Environment in relation to the sustainable use of the resources of Amazon Basin. Available in:<http://www.sema.mt.gov.br/index.php?option=com_content&view=article&id=414&itemid=346>. Access in: november 22nd, 2014.

MTE. Ministry of Labor and Employment. Order number 547 of 11th, march 2010. **It establishes, within the MTE, the Special Register of colonies Fish**. Available in: <http://www.sema.mt.gov.br/index.php?option=com_content&view=article&id=414&itemid=346>. Access in: november 22nd 2014.

NEW, M. B. Responsible aquaculture: is this a special challenge for developing countries? **World aquaculture-baton rouge**, v. 34, n. 3, p. 26-31, 2003.

NOGUEIRO, Luís A. S. **Practices of environmental management on the local public administration.** (Master's thesis). Department of Sciences and Food Engineering, New University in Lisboa, 2008, 136.

NORTHEAST BANK IN BRAZIL. **Environmental guide for farmers.** Fortaleza, Northeast Bank, 1999.

OSTRENSKY, Antonio; BORGHETTI, José Roberto; SOTO, Doris. **Aquaculture on Brazil**: the challenge is to grow. Brasília: Organization of United Nations for Agriculture and Food. 2008, 276p.

PELICICE, F.M. & AGOSTINHO, A.A. Fish fauna destruction after the introduction of nonnative predator (Cichla kelberi) in a Neotropical reservoir. **Biol. Inv.**, 11, 1789-1801, 2009.

PELICICE, Fernando Mayer et al. A serious new threat to Brazilian freshwater ecosystems: the naturalization of nonnative fish by decree. **Conservation Letters**, v. 7, n. 1, p. 55-60, 2014. Available in:< http://onlinelibrary.wiley.com/doi/10.1111/conl.12029/full>. Access in: february 20th 2015.doi:10.1111/conl.12029.

ROCHA, Carlos Magno Campos da et al. Advances on research and development of Brazilian aquaculture. **Brazilian Agricultural Research**, v. 48, n. 8, p. iv-vi, 2013. Available in: <http://www.scielo.br/scielo.php?script=sci_arttext&pid=s0100204x20130008 00003&len&nrm=iso>. Access in: december 15th 2013. http://dx.doi.org/10.1590/s0100204x2013000800iii.

ROUTLEDGE, E.A.B. ; ZANETTE, G.B. ; SALDANHA, E. C. L.; ROUBACH, Rodrigo. The importance of research for the development of the aquaculture productive chain. **Agricultural Vision**, v. 11, p. 4-8, 2013.

SAMPIERI, Roberto Hernández; FERNÁNDEZ-COLLADO, Carlos; LUCIO, Pilar Baptista. **Research Methodology**. Ciudad de México: McGraw-Hill, 2006. 850p.

SIMBERLOFF, D., Martin, J.L. & GENOVESI, P. Impacts of biological invasions: what's what and the way forward. **Trends Ecol. Evol.**, 28, 58-66, 2013.

SOTO, D., J. AGUILAR-MANJARREZ & N. ISHAMUNDA. Building an ecosystem approach to aquaculture. FAO Fish. **Aquacult. Proc.** Roma, 2008.   231   pp.   Available   in:< http://www.beijer.kva.se/ftp/wioaqua/faoeaa2007.pdf#page=23>. Access in: february 21th 2015.

SPSS. Statistical Package for the Social Sciences. **SPSS 11.5**: statistical algorithms. Chicago: SPSS, 2002.

STRAYER, D.L. Eight questions about invasions and ecosystem functioning. **Ecol. Lett.**, 15, 1199-1210, 2012.

UNITED NATIONS *(1992)* Convention on biological diversity. **United Nations**, Rio de Janeiro. Available in:<http://www.cbd.int/convention/text/>. Accessed in: february 18th 2015.

VEIGA, José E. Indicators of sustainability. **Advanced studies,** v. 24, n. 68, p.   39-52,   2010.   .   Available   in: <http://www.scielo.br/scielo.php?script=sci_arttext&pid=s0103-40142010000100006&lng=en&nrm=iso>. Access in: february 14th 2015. http://dx.doi.org/10.1590/s0103-40142010000100006.

VINATEA, L.A. A. **Aquaculture and sustainable development**. Editor UFSC, Brazil, 1999, 310 p.

VITULE, J.R.S., FREIRE, C.A. & SIMBERLOFF, D. Introduction of nonnative freshwater fish can certainly be bad. **Fish Fish.**, 10, 98-108, 2009. Available in:< http://onlinelibrary.wiley.com/enhanced/doi/10.1111/j.14672979.2008.00312.x /#survey. Access in: february 21th, 2015.

VITULE, J.R.S., Skóra, F. & Abilhoa, V. Homogenization of freshwater fish faunas after the elimination of a natural barrier by a dam in Neotropics. **Diversity Distrib.**, 18, 111-120, 2012.

WACKERNAGEL, Mathis; REES, William. **Our ecological footprint: reducing human impact on the earth**. New Society Publishers, 1998.

ANNEX 1 Questionnaire used on the research, Ariquemes and Colorado do Oeste, 2012.

	Name of the property____Municipality: ___Number of tanks___Total area (m^2 or hectare) of tanks:__ NA = Indecision of the interviewed			
	Indicators	YES	NO	NA
	1. Property rights and attention to the legislation			
1	Does the farm have environmental licensing of the property? (The environmental licensing of aquaculture in the Federal field has the IBAMA as responsible agency and obeys to the Resolutions CONAMA 237/1997 and 312/2002. In the State the licensing must obey the state legislation and cannot be more permissible than the Federal).			
2	Does the farm have legal documents prove the compliance with environmental laws for the fish farming on the property?			
	2. Relationship with the community			
3	Does the property interfere on areas of public use (areas of leisure that take advantage from the local potential in function of the localization and environmental characteristics)?			
4	If "yes" on the anterior question: are there many meetings to look for solutions for conflicts?			
5	Are there meetings with the community to discuss about the growth of pisciculture and other related issues?			
6	Does the farmer hire people that live next to the farm?			
7	Does the farm contribute to the welfare of the community (health, recreation and education)?			
	3. Workers safety and relationship with them			
8	Does the farmer know the value of the current minimum wage, including social charges?			
9	Are the wages in accordance with the legislation: minimum of R$ 724.00, vacation and retirement?			
10	Does the farm hire workers less than 18 years?			
11	Does the farm provide home to the workers, attending to the minimum: electrical and hydraulic installations, rooms, kitchen, bedrooms and bathrooms?			
12	Is the water offered to the workers potable?			
13	Does the farm provide food to the workers and respect the local mores of food consumption?			
14	Does the farm have the necessary for first aid (adhesive tape, gauze, anti-allergic, merthiolate)?			
15	Does the farm elaborate the emergence plan for serious accidents?			
16	Does the farm provide training to the employees about general safety, personal hygiene and first aid?			

17	Does the farm provide equipment of individual protection to the employees?			
	4. Conservation of protected areas			
18	Were the permanent preservation areas (riparian vegetation, hill top, floodplain) removed for the farm construction?			
19	In the case of "yes" on the anterior question: did the farm reduced the impact caused by the removal of the protected area (riparian vegetation, hill tops and floodplains)?			
20	Did the farmer reforest the property?			
	5. Management of the fishpond effluents			
21	Does the farm have records of the water and effluents monitoring?			
22	Does the farm protect channels and ravines to avoid erosion on the tank?			
23	Does the farmer minimize the water exchange on fish tanks?			
24	Does the farmer promote the natural productivity (plankton) to decrease the use of foods?			
25	Does the farmer drain the fishponds to avoid the suspension of plankton and algae undesirable to the fish farming?			
26	Does the farm use the effluents of water from tanks, redirecting them to the irrigation of plants?			
27	Does the farm dispose the bottom sediments of the tank to avoid erosion of fishponds?			
28	Are there aerators on the fattening tanks?			
29	Does the farmer perform chemical analysis of the fishpond water?			
30	Does the farmer evaluate the plankton of the fishpond to know the need of fertilization?			
31	Does the farmer know the water flow of the property used to complete tanks and fishponds?			
	6. Management of sediments			
32	Are the sediments removed from fishponds, weirs and other areas in the farm?			
33	In the case of sediments removal, are they deposited in an adequate place?			
34	Does the sediments removal cause soil degradation, with sand and soil loss?			
	7. Conservation of water and soil			
35	Were the fishpond and the farm built on permeable soil?			
36	Does the farmer adopt measures necessary to avoid the contamination of aquifers, lakes and other water bodies?			
37	Does the farm use chemical fertilizers and medicines within the fishponds?			

	8. *Species used*			
38	Does the farmer use exotic species (from rivers of other states and countries)?			
39	Has the farmer recorded the escape of exotic species to the natural environment?			
40	Does the farmer have structures that avoid the escape of exotic species?			
41	Does the property produce its own fries?			
	9. *Disposal of inputs and wastes*			
42	Does the farm store fuel, lubricant and agro-chemicals on appropriate locations?			
43	Are the fuel, lubricant and agro-chemicals next to the fish food?			
44	Does the farm deposit domestic wastes on closed containers protected from the water?			
45	Are the fuel labeled and put far from possible sparks and explosions?			
46	Does the farm comply with the laws regarding the management and treatment of wastes?			
	10. *Management of medicines and chemists*			
47	Does the farmer use antibiotics only after diseases diagnosis?			
48	Does the farmer use food with medicines for the fishes?			
49	Does the farmer prevent diseases by means of a good nutrition, correct management of fishponds and reduction of stress?			
50	Does the farmer know the list of medicines and chemists prohibited in Brazil and in the world?			
51	In the case of using antibiotics, does the farmer use the minimum necessary dosages?			
52	Does the farmer use ecological methods to control predators (birds and alligator)?			
	11. *Collection, processing and transport*			
53	Does the farmer verify the refrigeration temperature (4.4 °C) while transporting the fishes			
54	Does the farmer use sulfite to transport fishes? (Sulfites are antioxidant, conserving the fishes, and can cause severe allergic reactions).			
55	In the case of "yes" on the anterior question, is the solution of sulfite deactivated after its use?			
56	Are the workers protected to prevent infections when handling fishes?			
	12. *Traceability*			
57	Does the farmer have records of the fish treatments? (Sanitary state, productivity of the national pisciculture,. Represents concurrence and stimuli to the quality).			

Source: adapted from Carrasco (2006).

I want morebooks!

Buy your books fast and straightforward online - at one of the world's fastest growing online book stores! Environmentally sound due to Print-on-Demand technologies.

Buy your books online at

www.get-morebooks.com

Kaufen Sie Ihre Bücher schnell und unkompliziert online – auf einer der am schnellsten wachsenden Buchhandelsplattformen weltweit! Dank Print-On-Demand umwelt- und ressourcenschonend produziert.

Bücher schneller online kaufen

www.morebooks.de

OmniScriptum Marketing DEU GmbH
Heinrich-Böcking-Str. 6-8
D - 66121 Saarbrücken
Telefax: +49 681 93 81 567-9

info@omniscriptum.com
www.omniscriptum.com

MIX
Papier aus verantwortungsvollen Quellen
Paper from responsible sources
FSC® C105338

Printed by Books on Demand GmbH, Norderstedt / Germany